THE CHAMPIONS OF
CAMOUFLAGE

A FIREFLY BOOK

Published by Firefly Books Ltd. , 2019
Original Title: Les Champions du Camouflage
Authors: Jean-Philippe Noël & Biosphoto
Text © Editions Glénat 2018 – ALL RIGHTS RESERVED
Photographs © as listed on page 160

First printing

Library of Congress Control Number: 2019940486

Library and Archives Canada Cataloguing in Publication
A CIP record for this title is available from Library and Archives Canada

Published in Canada by
Firefly Books Ltd.
50 Staples Avenue, Unit 1
Richmond Hill, Ontario
L4B 0A7

Published in the United States by
Firefly Books (U.S.) Inc.
P.O. Box 1338, Ellicott Station
Buffalo, New York
14205

Translation: Travod International Ltd.

Printed in China

Jean-Philippe Noël

An animal lover since childhood, Jean-Philippe Noël is a journalist specializing in nature, travel and history. He is a regular contributor to various magazines, such as *Science & Vie* and *Trente millions d'amis*, and has written some 20 books, including children's books about animals. He is also the author of historical dramas for *France Inter* as well as theater productions.

Biosphoto

Biosphoto is a photo agency specializing in nature, the environment and gardens. It was founded in 1987 with an environmental advocacy mission. It has developed partnerships with NGOs, including Tara Expeditions. It now represents 450 photographers worldwide.

THE CHAMPIONS OF
CAMOUFLAGE

JEAN-PHILIPPE NOËL • BIOSPHOTO

FIREFLY BOOKS

CONTENTS

Surprised, the Eurasian bittern freezes, impassive almost invisible. Its yellowish-brown plumage, streaked and speckled with black, blends in perfectly with the golden surroundings of the reed bed. Standing among the reeds, it points its beak toward the sky, slimming its outline. The dark vertical stripes that run down its throat are the shadows of the tall stems. Those who have observed it say it can stay like that for hours; it is even said that it gently sways its body to the rhythm of the reeds as they are swept by the wind.

PREFACE

There's a small yellow spider, discreet and invisible, on a yellow flower. You have to look for a long time before spotting it. The goldenrod crab spider has made discretion a way of life. Unlike many of its arachnid counterparts, it does not weave a web but keeps on the lookout, motionless, waiting for a pollinating insect to come along to gorge on nectar. Using its first two pairs of legs, which are particularly long, it grabs its prey, injects it with venom and later absorbs the liquefied flesh of its victim, leaving only a dried cuticle behind. The spider seems to have foiled the vigilance of its prey simply by adopting the same color as the flower it is sitting on: yellow, white or even pink.

The goldenrod crab spider, capable of changing its hue over a few days, is one of the best-known examples of animal camouflage. The ability to change color is known as homochromy, which is camouflage based on the similarity of color between an animal and its environment, allowing it to hide from the eyes of its prey as well as its predators. This is the main role of camouflage: to eat and to avoid being eaten.

These biological imperatives have given rise to quite extraordinary, even incredible adaptations. In this book we discuss those adaptations while exploring spectacular images. However, camouflage is not only a complex concept, in some cases, it is also worthwhile to reconsider it.

Indeed, while studying the homochromy of an Australian crab spider, *Thomisus spectabilis*, the entomologist Astrid M. Heiling and her colleagues at Macquarie University made a surprising discovery: The flowers on which this spider is located are more frequently visited by pollinating insects than those not hosting any arachnids at all... While one might think that the spider is less visible to its prey due to adopting the same color as the flower, the researchers found that the prey in fact easily identifies the spider because of the ultraviolet rays it emits. As it turns out, among the seductive assets that flowers use to appeal to winged pollinators, ultraviolet light plays a leading role, and researchers are just beginning to understand how it works. According to the researchers, *Thomisus spectabilis* is far from hiding. Instead, it has developed a colored signal based on ultraviolet rays, which reinforces the attractive power of the flower.

This example illustrates the difficulty faced by those who study forms of homochromy and of camouflage more generally. The perceptions we have of nature and the way in which we interpret them are, in fact, only the product of our sensory system. Thus, since ultraviolet light is located beyond the spectrum visible to humans, we cannot understand how a bee sees flowers. The same goes for our perception of sounds, smells, vibrations, chemical signals, etc. We must therefore remember that the illusion provided by camouflage, whether it seems perfect or rather vague, only very rarely targets our sensory and cognitive systems. It is instead aimed at another living being's, which may be radically different from us.

That being said, let's explore the many and surprising aspects of camouflage in the animal kingdom — a world full of trickery and subterfuge.

The female crab spider, which is slightly larger than the male, has perfect homochromy. Depending on its surroundings, it can be yellow, white or even pink, and the change of color requires a period of several days. Lying in wait on its flower, the spider is waiting for its next victim: a butterfly, a fly, a bee... Any insect, even one much larger than itself, will do the trick. Once it has captured its prey the spider injects it with a venom that acts very quickly, since the bite often occurs near the victim's head.

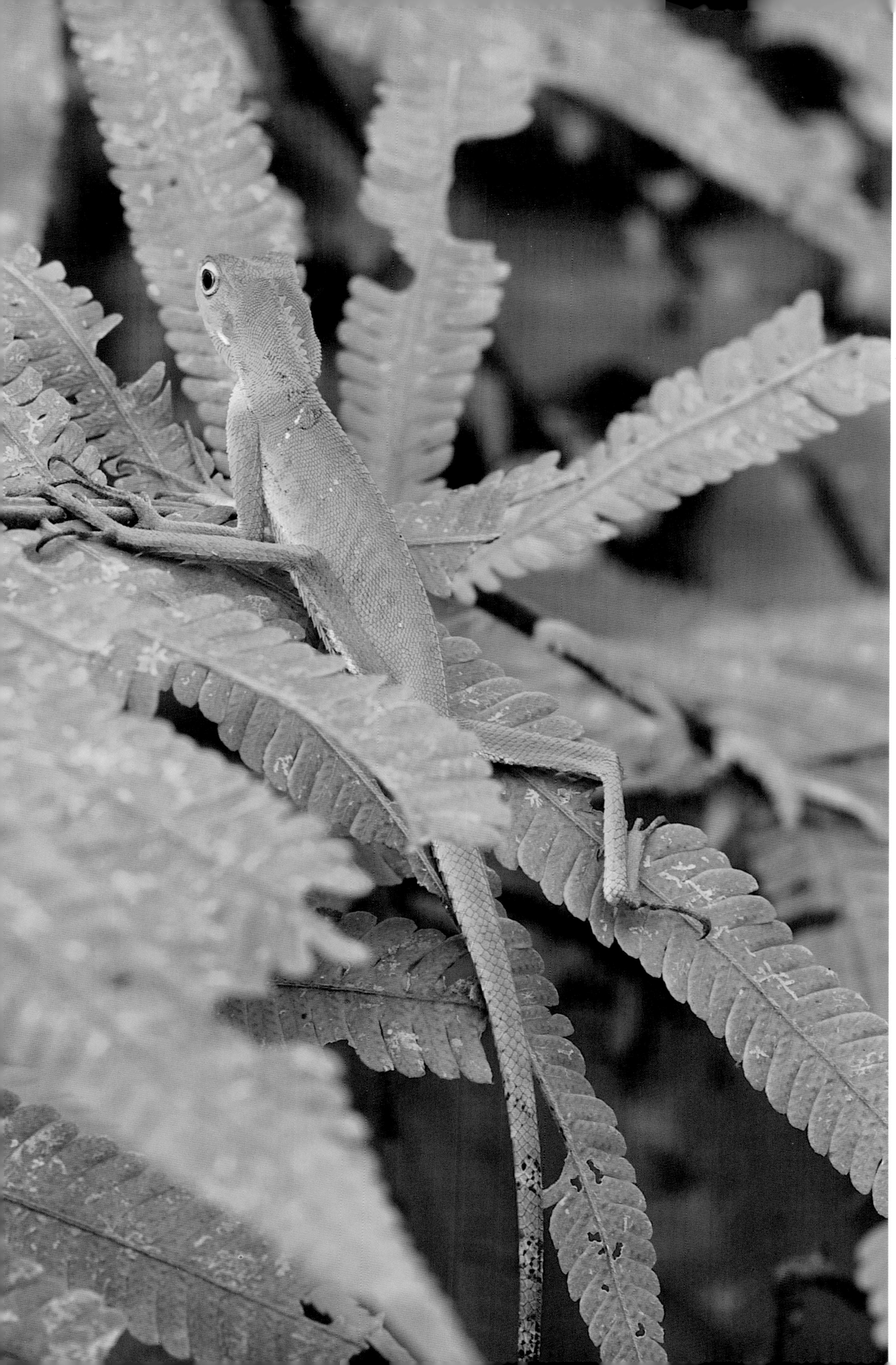

THE ART OF GOING UNSEEN

Eat and avoid being eaten! This fundamental law of nature has pushed evolution to invent ever more sophisticated survival techniques. Escape, venom, an unpleasant smell, concealment... Nature has no shortage of imagination in establishing a fragile balance between predators and their prey. Camouflage, or the art of going unnoticed, is one of the many strategies species use to ensure their longevity.

The most common form of camouflage in the animal kingdom is homochromy, which consists of showing colors similar to the surrounding environment. These are also referred to as cryptic colors (from the Greek *kruptos*, which means "hidden" or "secret"). Homochromy is found in all animal groups, from large green grasshoppers to polar bears, and comes in different varieties. It can be monochrome or multicolored. It is described as "seasonal" if the animal changes its appearance according to an annual cycle. It is called "variable" when the animal has the ability to change and reverse its coloring within a relatively short period of time. But animals do not like to be cataloged, systematized or confined, and the bridges between these types of homochromy are more numerous than scientific Cartesianism would allow.

Let's travel into a tone-on-tone world.

Previous spread
This beautiful lizard from Southeast Asia is a very bright, almost glowing green. The green crested lizard (*Bronchocela cristatella*) may also have a slight bluish tinge on its head. It measures 22 inches (57 cm), including its 17-inch (44 cm) tail. This excellent climber lives in forests, but it also frequents agricultural areas and sometimes gardens. When it feels threatened, it takes on a darker brownish-gray color.

Above
In this picture, a little frog clinging to lichen-covered bark is already almost invisible; in nature, you would have passed by it without even suspecting its presence. Active during the day, Webb's Madagascar frog (*Gephyromantis webbi*) lives on mossy stones along forest rivers. The coloring on its back, green with darker, irregular stripes, mixes with the hues of the lichen. Found only in northeastern Madagascar, around Antongil Bay, populations of this amphibian are currently declining, as it falls victim to deforestation, among other threats.

At right
The willow beauty is one of the most common moths in the Geometridae family. Present in most of Europe and Asia Minor, it inhabits a wide variety of environments. In Europe, several generations are born in succession between the months of April and November. The grayish-brown hues and designs on the wings vary from one individual to another. This moth is more or less visible depending on the surface its spends the day on. Camouflage is not always an exact science...

A CLOAK OF INVISIBILITY

Homochromy is the camouflage that allows an animal, whether prey or predator, to go unnoticed using only its body colors. Largely widespread in nature, one could almost state that it is the default camouflage. Homochromy can be relatively monochrome when the environment is itself composed of a dominant color; in this case, it is called "simple." It can also present in the form of polychrome patterns if the colors in nature are more mixed. In the context of simple homochromy, brown, with all its nuances, is among the most common colors for hair, feathers and scales. It could be called "all-purpose" or "neutral," as it doesn't attract the eye and blends naturally into the environment. A red deer can stand perfectly motionless in the shadows of a forest, and a wild boar can go unnoticed in a thicket of thorns even at a short distance from a hiker. The same goes for smaller animals. The brown hare, the wild rabbit and the alpine marmot all have a relatively solid-colored coat, making them look at first glance like a clump of earth or even a pile of scorched grass — in any event, something that would not be of interest to a predator. The same principle applies to many xylophagous insects. It is not surprising to find wood-colored cuticles in different species of wood-boring weevils, furniture beetles and brown lyctus beetles, an omnipresent species well known to woodworkers, cabinetmakers and carpenters.

The other "all-purpose" color, since it is ubiquitous in nature, is the green of chlorophyll, the main pigment for photosynthesis. Suffice it to say that from the ground to the treetops and through the first few feet underwater, every imaginable shade of green is present. It is adopted by animals whose size allows them to hide in vegetation. We find a multitude of arthropods, including the green huntsman spider, which is native to Northern and Central Europe, the large green grasshopper and the European lantern fly, a small insect related to cicadas. Many reptiles, such as the western green lizard, a large lizard found in temperate regions, sport green scales. In the tropics, where trees are never entirely bare, green is adopted by arboreal snakes, such as the green mamba, emerald tree boa and green vine snake, which spend the majority of their life among the branches, hunting from a hiding place. In such foliage, they cross paths with different species of birds that have predominantly green plumage, including several species in the Psittacidae family, such as the rose-ringed parakeet and the vernal hanging parrot, as well as many hummingbirds.

More monochrome environments also encourage homochromy. In hot, sandy desert regions, animals often opt for a buckskin coloring. This is easily observed in the fennec, whose color varies from pale brown to light beige, the sand cat and the desert jerboa. The vast expanses of the polar regions, where snow is present all or part of the year, have encouraged the whiteness of the polar bear and the ivory gull. The gyrfalcon is a fine example of adaptation. In temperate regions, the species has dark brown or gray plumage, and the farther north they are found, the lighter their feathers are. In Greenland, one can find the so-called "white" form, with an almost immaculate white underside. Other colors are rarer, but yellows, reds and oranges can be found among reef inhabitants, such as frogfish and shrimp.

Homochromy can also be composed of shades and tones that are more or less mixed, spotted or striped. The animal sports coloring in harmony with a particular environment. This is the case with many species

Thanks to its green color and the spots on its wings, every detail of this katydid photographed in Borneo blends in with the moss- and lichen-covered leaf on which it is holding still. The way it keeps its wings flattened and its forelegs forward heightens its excellent camouflage.

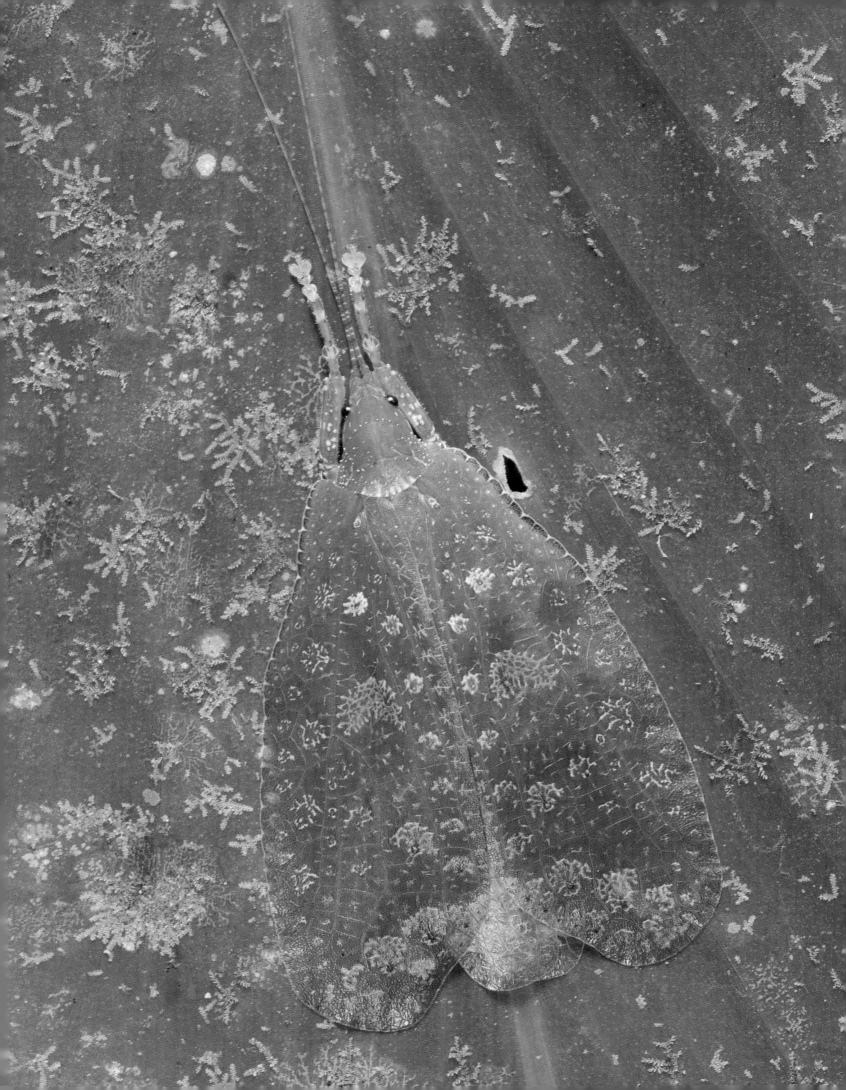

of amphibians, insects and female ground-nesting birds, along with their chicks and eggs.

Homochromy, however, does not necessarily mean uniformity. The color tones of animals are not fixed; they can differ from one individual to another and according to different environmental factors. Many mammals have coats that are darker or lighter in color depending on the season. Likewise, some frogfish change color in order to fit in better with the sponges surrounding them. In these fish, the changes can take from several days to a few weeks. Such changes in color should not be confused with seasonal homochromy (see page 36), whereby the color of the entire coat is modified, or with variable homochromy (see pages 42–43), which involves rapid changes.

Homochromy also has its limits. It cannot adapt to a frequent change of environment. In the great plains of the Serengeti during the dry season, the fairness of the lion can easily hide its presence. Hidden in the middle of tall grasses yellowed by a lack of water, the feline is likely to surprise a zebra. On the other hand, it is much more difficult for it to blend in with a carpet of young green shoots during the rainy season. Moreover, prey and predators have a mental picture of each other. The big eyes of the gazelle sweep the horizon in search of a known and feared silhouette, while the chickadee hopes to spot a butterfly thanks to the characteristic shape of its wings or from a detail of its anatomy, like a pair of antennae.

To improve their homochromy, many species have added disruptive patterns. As suggested by the term "disruptive," which comes from the Latin *disrumpere* (to break, smash into pieces), these patterns consist of "breaking" the silhouette of the animal through contrasting tones and designs. Thus, the black lines on tigers and the spots on panthers and cheetahs form irregular and fragmented silhouettes, which are difficult to detect in an environment of dense vegetation. In various species of plovers, such as the common ringed plover and the Kentish plover, the brown hues on their back merge with the soil and pebbles on beaches, while the contrasting shades (black, white and brown) on their head, neck and breast break up their silhouette. For these birds, which nest in open ground, on the beach or among pebbles, discretion is a matter of survival. As long as they don't move, they are almost invisible, even at a few feet away from the observer. Studies carried out with fake butterflies with disruptive coloration have shown that their chances of "survival" are better than those of animals with simple homochromic color.

Whatever the type of homochromy, it is often accompanied by behavior that enhances the animal's camouflage abilities. First and foremost there is immobility, which is camouflage's essential corollary. At the slightest warning, young ostriches, with their perfectly cryptic down, literally ground themselves in the grass, staying frozen until the danger disappears, and even the most perfectly camouflaged insect would be spotted by the slightest movement of its wings or antennae. It is also known that many animals are able to select the areas where their camouflage is most effective. Quails from Japan choose the color of the substrate that will provide the best protection for their eggs. Similarly, studies on moths by South Korean researcher Changku Kang, of Mokpo National University, have shown that they are able to orient their position at rest, with their wings open on a trunk, so that their wing patterns match their background environment as closely as possible.

Transparency, a form of camouflage close to homochromy, is used by many living beings, since it involves both the cornea of the eye and the wings of many insects. When all or part of an animal is transparent, the transparency provides an almost

perfect camouflage. Thus, some species of small
shrimp from the genus *Periclimenes* are as well concealed as Harry Potter under his invisibility cloak.
These shrimp live symbiotically on another animal,
usually a sea anemone.

In an aquatic environment, simple homochromy is
replaced in several species of fish and marine mammals by countershading. It is no longer a matter of
getting lost in the dominant colors of the environment but of playing with light intensity. Rays, sharks,

mackerel, herring and many cetaceans have dark backs and a white, cream or light gray belly, which reduces the contrast between the animal's body and its environment. From above, the dark silhouette disappears in the blue depths; seen from below, the clear tint of the belly is lost more easily in the light filtered by surface water. Countershading is also found in all penguin species, as well as in many diving birds, such as guillemots, little auks and puffins.

One cannot close a chapter on homochromy without mentioning a case beloved by evolutionists: the pepper and salt geometer, also known as the peppered moth. This small moth, which during the day rests on the bark of birch trees, exists in two types: the *typica* form, whose white wings are streaked with black, and the *carbonaria* form, which has larger black marks. While the latter had long been the rarer variety in England, it became far more widespread after the industrial revolution. It's believed that as fumes from factories blackened tree bark, the moths adjusted their coloring accordingly. What surprised scientists is that it only took a few decades. For example, in the Manchester area, the *carbonaria* form was still very

rare in the 1850s; a century later, it represented 98% of the population. Since the 1960s, the reduction of emissions and other pollutants has allowed birches to regain the light color of their bark – and the moths have regained the whiteness of their wings.

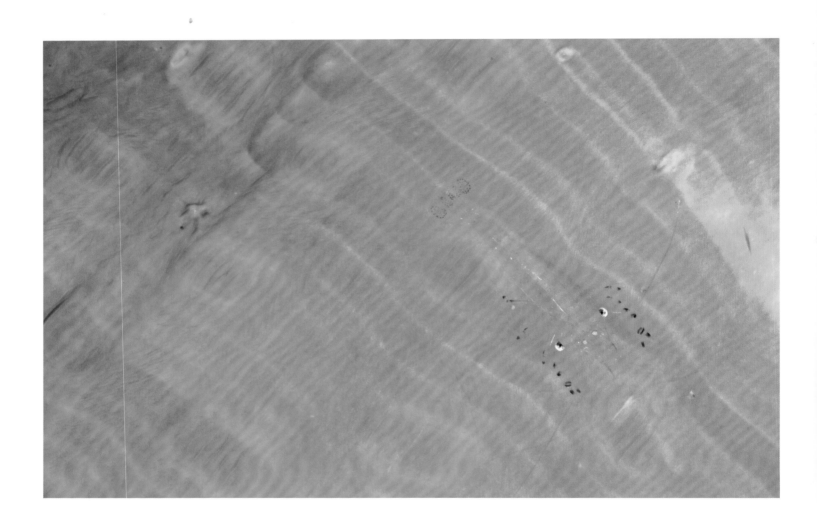

Body transparency is a camouflage strategy developed among many marine species. It allows an animal, like this shrimp from the Philippines, to disappear in any environment. In optical terms, transparency is the ability of a material not to reflect the light it receives. The light passes through the body. The transparency doesn't have to be a permanent state. Thus, depending on their state of stress or activity, some shrimp species become more opaque, regaining their transparency once calm is restored.

As its name indicates, the crinoid shrimp lives on... crinoids, which are marine animals that are close relatives of the starfish and sea urchin. The color tones of this shrimp, which measures between ¾ to 1 inch (2–3 cm) can vary from one individual to the next, ranging from yellow to white or even brown, but they are always slightly streaked with lighter bands. Hidden in the arms of its host, it feeds on zooplankton and the small debris that it filters. It is found in the Coral Sea, along the northern coast of Australia as well as in Papua New Guinea, and as far north as the Philippine Sea and as far east as New Caledonia.

In nature, there are few species with an extremely bright homochrome color. The painted frogfish, however, is one example. These fish often live among sponges and eventually adopt a similar coloring. The above specimen has black eyes that are reminiscent of the holes found in the structure of the sponge it is resting on. The lighter spots are actually sand, which sticks to the skin of the fish, probably due to its mucus, further enhancing its camouflage.

The color changes of the frogfish can be radical, going from deepest black to bright red. During this process, it passes through different phases of coloration, sometimes very far from the beginning and ending colors. Observed in an aquarium, these changes generally happen over several days and up to a few weeks.

Above

Can you find the animal that is hiding in this picture? It's a small moth. This moth, known as Barrett's marbled coronet, is found across much of Europe. These moths have been observed placing their body in such a way that the pattern on their wings matches whatever they are resting on.

At right

Perfectly mimetic on the tree trunks they inhabit, tree frogs of the *Dendropsophus melanargyreus* species are restricted to South American tropical forests. Their brown, gray, red or beige limbs form a blotchy patchwork, sometimes ringed with black. Large, lighter spots, often beige, break up the contours of their silhouette. This species is a perfect example of homochromy with disruptive patterns.

Previous spread, at left
Perfectly invisible against the trunk it is clinging to, this gecko's camouflage mimics lichen to perfection. Homochromy is not its only trick for avoiding predators, however. Once deployed, the skin folds along its body as well as its widely webbed head and fingers allow gecko species of the genus *Ptychozoon* (here, *Phychozoon rhacophorus*) to escape by gliding in the air, as if they were equipped with a parachute. These geckos are found in Southeast Asia and India.

Previous spread, at right
Originally from Madagascar, this large gecko, sometimes called the "leaf-tail gecko" and measuring about 6 inches (15 cm) long, relies on the perfection of its camouflage to go unnoticed. Once pressed against the bark, the shape of its body disappears entirely. Its camouflage is reinforced by a strip of skin running along its body that avoids any projection of a shadow. The animal remains motionless all day, waiting until nighttime to hunt.

Above
The southern rock agama, a species of gecko, lives in Southern Africa, where it is still quite common. This gecko's ability to change color is not as marked as the chameleon's, but the male can change the color of its head from blue, which it sports during the mating season, to a less conspicuous shade. The female maintains a relatively cryptic coloration all year round.

At right
This strange animal — flattened against a leaf and with a fringe that perfectly hides the contours of its body — is a larva. As adults, the species of the Ascalaphidae family vaguely resemble dragonflies. Some African species have homochromic larvae, which is camouflage that is both defensive and offensive in nature, since these larvae are formidable predators who use their powerful mandibles (seen at the bottom of the photo) to seize insects that pass within their reach.

At right
The large family of wolf spiders (Lycosidae) do not weave webs but hunt on the ground. Their generally neutral coloring allows them to be perfectly discreet. Wolf spiders are found on every continent except the poles.

Following spread
Spiders of the genus *Cryptothele* are only known by a few rare specimens that represent a dozen species found in Asia and Oceania, and in half of these cases, only the females have been scientifically recorded. Evidence that good camouflage can even deceive researchers!

SEASONAL OUTFITS

Some animals have different camouflage depending on the season; this is called seasonal homochromy. This phenomenon is found in ermine, mountain and Arctic hares, rock and willow ptarmigans, and the polar fox, all of which live in regions that experience heavy snows in winter and have plumage or coats that change from a brownish-red, earthen color to a more or less immaculate white that makes them less visible in the snow. This transformation, which occurs through molting, often requires several weeks.

Let's follow, for example, a year in the life of a rock ptarmigan. This Arctic bird settled in Western Europe during glacial periods and survived only in mountainous regions during the warming that followed. Depending on the season, it lives at altitudes of between 5,900 to 9,850 feet (1,800–3,000 m), with a clear preference for northern slopes. During the course of a year, the bird changes its plumage three times. During the winter, from the end of November to the end of March, the male and female are white, with the exception of their black tails. In that season, they frequent the wind-swept summits, or if food is lacking, they go down to the forest, gorging themselves on the buds of willows and rhododendrons. On the other hand, they always spend the night in snow-covered areas, lodging themselves in a hollow dug out of the snow, which, like an igloo, helps them conserve their body heat.

From April to October, males display a predominantly dark gray plumage, while the hens are a tawny brown. Both sexes are more or less variegated, with thin black, white or brown stripes. This plumage is perfectly discreet for the hens, which, during this time, nest on the ground in areas of low vegetation and among rocks. In autumn, the ptarmigan's plumage is grayish and flecked with white on the top, while the underside remains whitish.

It should be noted that males take longer than females to acquire their spring plumage. As a result, when the snow is already starting to melt, they are more easily spotted and are therefore more heavily predated by raptors.

In the ptarmigan, ermine and polar fox, the color is due to the presence of a well-known pigment, melanin. This is synthesized thanks to the action of melanocyte-stimulating hormones, whose activation depends on the rate of ultraviolet radiation. These hormones are therefore much more active in summer than in winter, which leads to an increase in the amount of melanin synthesized in the animal's integument during the warmer months. Conversely, during the colder months, when the level of ultraviolet rays decreases, the amount of melanin synthesized decreases, and the fur and plumage gradually whiten.

There are also cases of seasonal homochromy in insects. This is the case, for example, with the green lacewing, also called the "golden-eyed fly," which is much appreciated by gardeners because its larva is a great eater of aphids. While in summer it is a shade of green that allows it to blend in with vegetation, it turns brown at the end of the season to become more discreet during the winter, which it usually spends in a dark nook of a house. The green shield bug, also known as the green stink bug, is certainly better known for its nauseating odor than for its change of appearance. As the name suggests, it is green in summer, when it is treating itself to plants, but it becomes brown in the fall, while preparing to find a quiet place to spend the winter. The shield bug and the lacewing will don their green outfits and find their way back to the garden again in the spring.

Like its cousin the rock ptarmigan, the willow ptarmigan changes its plumage three times over the course of a year (not twice, like the majority of birds). Essentially white in winter, the willow ptarmigan is brown with many shades of red, gold and gray in the summer. Its mid-season appearance is a mixture of both. This molting allows it to have a plumage that is always suited to its immediate environs. It frequents tundra and boreal forests with lakes over a large area of the northern hemisphere.

Above

On the island of Spitzbergen in Norway, the silhouette of a polar fox, or Arctic fox, is barely discernible among the boulders. Native to Arctic regions, this compact fox generally weighs less than 11 pounds (5 kg) and has thick fur that effectively protects it from extreme cold. It is also the perfect camouflage. Brownish-red in summer, the polar fox gets an entirely white coat in winter. But not all of them show this change in coloration; there is a so-called "blue" form. The fur color of these foxes varies in summer between black and brown and changes to a grayer shade (called "blue") in winter.

At left

In the summer, the rock ptarmigan lives in open landscapes with low, sparse vegetation. Males choose territories along rocky outcrops that allow them to monitor the females. Found in high mountains, more than 6,500 feet (2,000 m) above sea level, the species is threatened by global warming, which in the longer or shorter term may alter the conditions necessary for breeding periods and could fragment local populations.

Following spread

Immobile in the immensity of a field of snow, the mountain hare, also called the "snow hare," is barely visible. Like the rock ptarmigan, the mountain hare is a survivor from the ice age. It is found from Finland to Siberia, and some populations also live in the United Kingdom, Ireland, Poland, France and Japan. While white in winter, these hares get a gray coat for the summer, beginning in May. The animal's molting is the result of a hormonal phenomenon that is triggered by the warmer temperatures and longer daylight hours.

EXPRESSIVE COLORS

The ability of chameleons to change color quickly is legendary. In his *History of animals* (c. 350 B.C.E.), Aristotle explains: "The change in the chameleon's color takes place when it is inflated with air. It is sometimes black, not unlike the crocodile, or green like the lizard, but streaked with blank like the panther." Chameleons (of which there are more than 200 species) are probably the best known example of variable homochromy in the animal kingdom. Variable homochromy involves, as you may recall, an animal changing the color of its integument within a short space of time, along with the ability to change it back.

What makes these spectacular costume changes possible is the presence, in the superficial tissues of the dermis, of pigment cells called "chromatophores." These form the basis of the coloring of the animal's skin and allow it to increase or reduce the amount of pigments in its integument.

There are several types of chromatophores. Melanophores are responsible for browns and black; these are the most active chromatophores during color changes. Lipophores are the basis for yellow, orange and red colors. In rare species of fish there are cyanophores, which contain blue pigments. Leukophores and iridophores are chromatophores whose function is to reflect light. These are the cells that create the iridescent and metallic colors of butterfly wings, the silvery skins of fish and the shiny reflections of many feathers.

The color changes are either morphological or physiological. For the former, it is the amount of pigment in the cells that varies. This type of change, which occurs slowly, mainly concerns seasonal homochromy. Rapid and reversible, the physiological color changes are related to the dispersion of pigments on larger or smaller surfaces, particularly by the extension or contraction of the chromatophores. Physiological color changes are therefore related to variable homochromy. The latter is mainly found in amphibians, snakes, lizards, bony fish, cephalopods and crustaceans. In order to quickly adapt their color to their environment, animals capable of variable homochromy must perceive the different colors that surround them. To do so, they, of course, use their eyesight. But some species also have cutaneous photoreceptors directly associated with the chromatophores. These photoreceptors react more specifically to changes in light and temperature. Depending on the species, the chromatic variations are controlled by hormones or neurotransmitters, or sometimes both.

Below and at left
Comparing these two images, we understand how the chameleon's ability to change color has become legendary... In Madagascar, the panther chameleon has an incredible palette of colors, which is reserved mainly for the male, with the females being more drab. Within the same species, the colorations differ according to the location. Thus, those on the island of Nosy Be tend toward blue and green tones, whereas on the northwest coast, pink and yellow dominate. These colors are embellished by orange, red and black tones in the form of bands, spots and stripes.

Brookesia stumpffi is a small chameleon less than 4 inches (10 cm) long. It lives in northeastern Madagascar, on the ground or on low branches. It looks like the dead leaves that it hides among. Its base colors are brown, gray and a dull green. Although the male during the mating season takes on an appearance similar to lichen, the ability of this species to change colors is less dramatic than that of other chameleons, as its color spectrum is less extensive.

In addition to being able to change their colors "at will," some species have adaptations of different shades depending on the type of predator that hunts them and the visual abilities of those predators.

A study conducted on *Bradypodion taeniabronchum*, a dwarf chameleon native to the Eastern Cape in South Africa, showed that, when stationed on a branch, it reacts differently depending on whether the predator comes from the sky or from the ground. If the predator is a bird, the chameleon densifies its color in order to blend better into the vegetation; on the other hand, when dealing with a snake, which climbs up from the ground, it brightens its color tones to drown them in the light of the sky, in a way using the principle of countershading (see pages 14–15). These changes can even account for the visual abilities of its different types of adversaries.

Another champion in terms of variable homochromy is the common cuttlefish, which adapts its camouflage to perfectly match the substrate on which it finds itself: On sandy soil, it has a uniform color; on a bed of gravel, it has a speckled pattern; and on stony or rocky ground, with greater contrasts, it adopts disruptive patterns. In all cases, it is ideally concealed. Able to anticipate while

in motion, the cuttlefish begins to change its colors even as it approaches the new environment it is heading toward. It disappears all the better at the sight of its predator, changing its color and even its shape. This change of costume and appearance takes less than a second and is one of the fastest reactions in the animal world.

Previous spread
Disappearing in a few seconds, the tropical flounder — like other flounders (genus *Bothus*) — adapts its body color to the sea bottom below it, thus avoiding being spotted both by its prey and by its predators. Studies from 2015 show that, in order to camouflage itself, a flounder chooses environments with fairly neutral colors, such as sandy bottoms, or that are covered in dead coral debris. These environments match their chromatic scale, and they find them by sight. The tropical flounder is generally found along sandy bottoms and coral reefs in shallow waters (less than 500 feet [150 m] deep) of the Indo-Pacific basin.

Above
Invisible on sand, gobies (family: Gobiidae) are small fish that live on sandy seabeds. As they are poor swimmers, they never move away from the bottom. Although some gobies sport very showy colors, other species are, once stationed on the ground, completely invisible. Some are also able to change color.

At right
True quick change artists of the seas, cuttlefish are not content with a simple change of color. To perfect their camouflage, they are also able to transform the texture of their skin. Some, like this broadclub cuttlefish, can also change shape. It can place two of its tentacles so as to resemble branches of coral as it lies in wait.

Note that variable homochromy is not only used for camouflage. Chameleons can use it to express their "state of mind," with bright colors betraying sexual or aggressive behavior. Some animals, such as the bearded dragon, also use it for thermoregulation. This Australian lizard lightens or darkens its skin depending on the outside temperature. In cold and wet conditions, the African frog *Chiromantis xerampelina* is darker than on a hot, dry day. The same goes for some ghost crabs (of the *Ocypode* genus), which are lighter during the day than at night. This allows them not only to limit their absorption of the sun's rays during the day but also to go unnoticed at night. Variable homochromy can be a virtual Swiss army knife, which animals can use, depending on the circumstances, to hide, regulate their body temperature or communicate.

The cuttlefish's ability to match its environment is due to the three optical components stacked on top of each other in their integument. Leucophores reflect light, iridophores create iridescent colors and chromatophores contain different pigments. It is by dilating or contracting these cells that they can produce different color tones and patterns.

Not only is *Abdopus aculeatus*, a small octopus found in the Indian Ocean, able to melt completely into its environment, it also displays mimicry when it needs to make a quick escape. Unlike the majority of its congeners, it does not escape by swimming but by "walking" on two of its tentacles. This behavior makes it look like algae floating in the water and confuses most of its predators.

GETTING INTO SHAPE

The art of camouflage is not just about taking on the colors of the immediate environment. Some species can also imitate shapes, making themselves look like a natural element of their surroundings. This type of camouflage is called "homotypy." In this case, the animal can be seen perfectly well without revealing its true identity. There are "counterfeits" of leaves, branches, twigs, pebbles, even animal or plant species... Some grasshoppers mimicking leaves even go so far as to simulate the nibbling of herbivores or the presence of mold. Some homotypies are accompanied by behavioral mimicry, such as the quietly balanced movements of a stick insect, comparable to the shivers of a leaf in the wind.

Nature's capacity to copy is always a source of fascination for naturalists. Since Darwin and the early theories on mimicry were developed by the entomologist Henry Walter Bates in the 19th century, homotypy has been considered fundamental evidence of the theory of evolution. Nevertheless, the mechanisms that allow such mimicry between living beings that belong to different kingdoms are still poorly known, even misunderstood.

Previous spread
Called dead leaf mantises, the *Deroplatys* genus are native to Asia and live in forests close to the ground, where their appearance, reminiscent of dry leaves, hides them from the sight of their predators and future victims alike. Depending on the species, their size varies from 1³/₄ to 3 inches (45–80 mm) long. Females, always more corpulent than males, are unable to fly. When disturbed, the mantis abruptly spreads its wings, revealing bright colors often embellished with ocelli.

Opposite page
The large belly of this male leaves us in no doubt: This lined seahorse is taking care of its future offspring. Perfectly concealed thanks to its dull color and cutaneous growths that mimic aquatic plants, it is waiting for a happy event. In seahorses, the female lays her eggs in the ventral pocket of the male, who holds them until they hatch. This large seahorse, nearly 8 inches (20 cm) long, is, like all seahorses, threatened by commercial fishing, pollution and the loss of its habitat.

At right
Thanks to its long body, this ogrefaced spider (genus *Deinopis*, also known as net-casting spiders and gladiator spiders) is perfectly hidden among the blades of grass. To capture its prey, it uses a net that it holds between its legs and projects onto its victims. Hence the alternate names net-casting spider and gladiator spider, the latter a reference to the famous combatants of the Roman arena, some of whom were armed with nets..

LIKE LEAVES

The big island of Madagascar has many endemic species. Among them is one of the most unique groups of geckos. Living only in the humid Malagasy forests, the genus *Uroplatus* consists of about 15 species. They are characterized by their large eyes, triangular head, flat tail and especially by their ability to perfectly blend in with nature. While some, such as the mossy leaf-tailed gecko (see page 29), perfectly mimic the colors of the bark of the trees they cling to, others, such as the satanic leaf-tailed gecko, disguise themselves as a dead leaf. Body and tail flattened, brown hues sometimes tinted toward yellow or orange, outlines similar to veins... The illusion is flawless during the day, when the animal stands motionless, confident in its camouflage. Like other *Uroplatus* geckos, it only becomes active after dark.

Terrestrial vertebrates mimicking leaves are rare. This homotypy is found in *Uroplatus ebenaui*, a close relative of the satanic gecko endemic to northern Madagascar, and in amphibians of the genus *Megophrys*. These small Asian frogs live on the forest floor, where their brown and black hues and the broken lines on their back evoke the tough, dried leaves of many tropical trees.

Numerous insects have successfully adopted a leaf theme for their mimicry. Notable members of this group are the Phylliidae, also called "leaf insects," found in Asia and Australia, and many species of mantis, like those of the genus *Deroplatys*, from Southeast Asia, and *Phyllocrania*, from Africa, including Madagascar. This type of camouflage is also very common in Orthoptera

This satanic leaf-tailed gecko brings the art of disguising itself as a dead leaf to the extreme. How? Through its colors, which vary in individual specimens from brown to yellow or almost black; its veining, which copies that of leaves; and the shape of its body and head, which are relatively flat. The overall look reaches its peak with the shape of its tail, which, in addition to having a central vein, bears traces of nibbling from herbivorous insects. This gecko, which is about 6 inches (15 cm) long and originates from Madagascar, remains motionless throughout the day in the branches of trees.

At left

Belonging to the genus *Megophrys*, the Malayan horned frog, also known as the long-nosed horned frog, lives in Southeast Asia. It owes its name to the two extensions that are above its eyes. It is especially remarkable for its ability to blend into piles of dead leaves. This camouflage is as defensive as it is offensive. As it sits motionless, the frog is not easily identifiable by predators, but it also hunts by lying in wait, watching for snails, worms, large arthropods and even other frogs, which it catches with a flick of its tongue.

Following spread

It's hard to tell the difference between the original and the copy! The insects of the genus *Phyllium* (family Phylliidae) have pushed the leafy copy to perfection. The young, which hatch on the ground among dead leaves, are a brown color. They take on their green color after about 10 days, when they join the leaves of the trees and shrubs they feed on. As adults, they measure about 4 inches (10 cm) long. These species live in South Asia.

(an order of insects that includes locusts, grasshoppers and crickets). For example, grasshoppers of the genus *Typophyllum*, which includes about 40 species, have made it a way of life. They look like ultra-realistic green or dead leaves, sometimes even including predation by insects or infections by microscopic fungi. This is an art that the Orthoptera have had a lot of time to refine. Indeed, 270 million years ago, some grasshoppers already looked like leaves. This was revealed by the analysis of a fossil discovered in the Alpes-Maritimes region of France. This study, conducted by French researchers in 2016, allowed us not only to discover the oldest known grasshopper, but also to identify the oldest case of homotypy observed to date.

Above
Among the dead leaf mantis (genus *Deroplatys*), which includes 12 species, the *Deroplatys dessica* is the largest, at just over 3 inches (80 mm) long with a wingspan of 3½ inches (87 mm) in the females. The males do not exceed 3 inches (75 mm) long, but their wingspan reaches 4¾ inches (120 mm) which allows them to fly. This species is found in Malaysia, Sumatra, Java and Borneo, where, like the other *Deroplatys*, it favors tropical forest habitats.

At right
It's not easy to see where the head and the tail are... In the vast order of Orthoptera (grasshoppers, locusts and crickets), there are many leaf-copying species, among which is the genus *Chorotypus*, shown here, which is found in Asia and Africa. There are nine species, including one discovered in Vietnam in 2017. They are all characterized by a laterally compressed body, a hypertrophy of the prothorax — which partially covers the head and often a part of the abdomen — a particular wing shape and an enlargement of the femurs, all of which gives them the appearance of a small dead leaf.

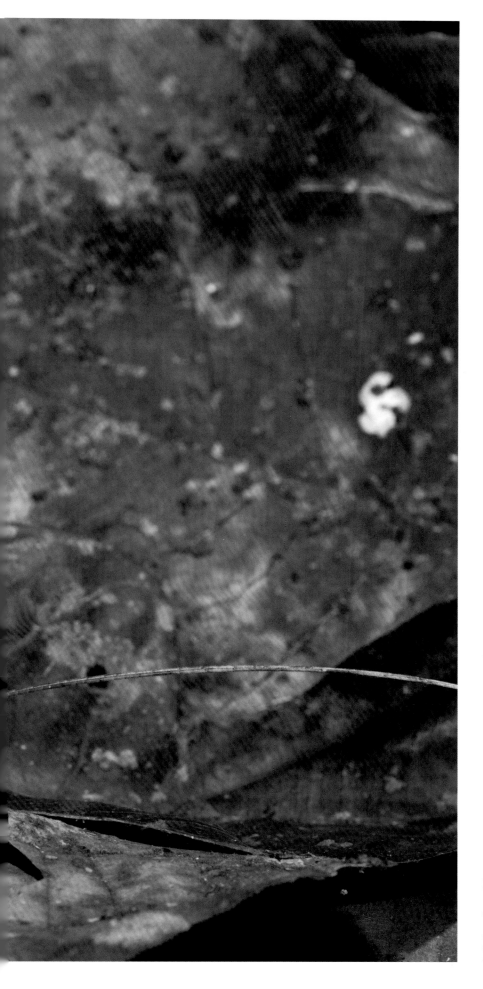

Although much younger, the oldest known butterfly is 190 million years old – the lepidopterans are not to be outdone. Many species mimic leaves, such as the common brimstone (*Gonepteryx rhamni*) and the lappet (*Gastropacha quercifolia*), which mimics an oak leaf.

How could evolution have given rise to such sophisticated copies? In an attempt to provide an answer, Takao Suzuki and his colleagues at the National Institute of Agrobiological Sciences in Japan were interested in an Asian butterfly genus, *Kallima*, whose perfect imitation of a leaf had astounded Alfred Russel Wallace in the 19th century. The British naturalist described it as "the most wonderful case of defensive mimicry in a butterfly." He also noted that they all look different, even within the same species, so as not to allow predators to become accustomed to a single appearance. By working on two species, *Kallima inachis* and *Kallima paralekta*, and 45 related species that are not necessarily mimetic, the researchers were able to demonstrate that they all shared a basic model that gradually evolved to give each one its current appearance. By placing these species on a genealogical tree developed through molecular genetics, they were able to trace the evolution of the basic pattern to the cryptic form. It is by progressing in small steps from this base that the two species of *Kallima* studied have developed morphological characteristics that, when they stand still and close their wings, make them look like perfect leaves.

Another great genus of leafy grasshopper is the *Typophyllum*. In these species, sometimes referred to as "little walking leaves," it is the limb-shaped wings that are made into leaves. Their green and brown hues have indentations that mimic mold or nibbling by insects and lines that draw veins like those on a leaf. In a species discovered in 2017, *Typophyllum spurioculis*, the male's wings also serve as a sounding board during its singing, which is unusual among the Orthoptera.

Previous spread
It is in a wide shot such as this one that we can truly appreciate the perfect mimicry of certain insects. As long as this katydid (family Tettigoniidae) does not move, it is almost impossible to notice it posing on its branch. Standing upside down, its central vein and the tip formed by its wings mimic the structure of a leaf.

With its wings closed, this comma butterfly looks like a damaged dead leaf. When open, its wings are orange and dotted with brown spots. Every summer, two generations follow one another, the second much darker than the first. This second generation does not die in the fall but finds a quiet place for the winter. Researchers speculate that these darker tones could be used as camouflage for butterflies that spend the cold season in the shelter of a trunk or branch.

The French name of the lappet strongly suggests what it mimics: *Feuille morte de chêne* means "dead oak leaf," which is exactly what this moth looks like. But is this really camouflage? There is some doubt, since this species flies mainly in July and August, when oak leaves are far from the mahogany color of its wings. In this large moth, found in most of Europe, females can reach a wingspan of 3½ inches (90 mm). Well camouflaged or not, the lappet, which lives on fruit trees, suffers terribly from the use of insecticides.

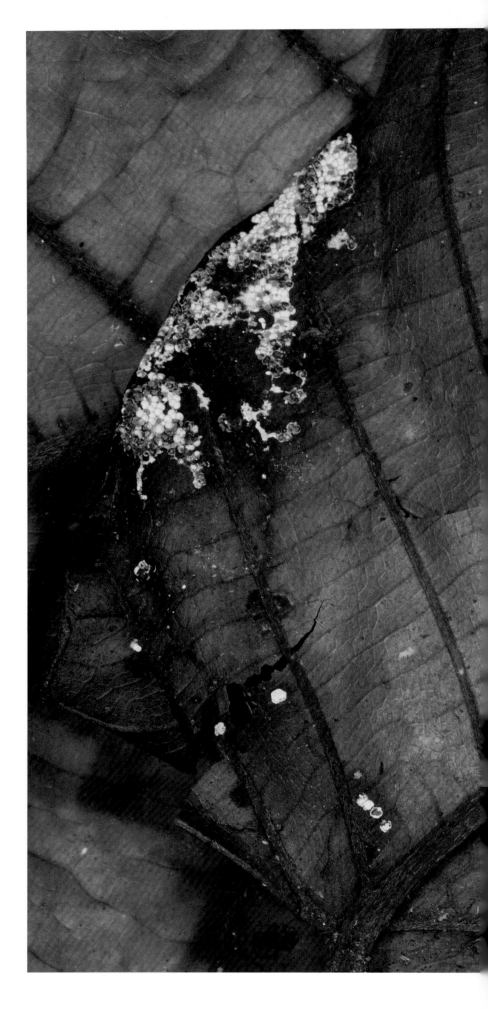

At right

It is while walking in the forests of Central and South America that one might hope to cross the discreet butterflies of the *Cimicodes* genus (here, *Cimicodes albicosta*). Their mahogany brown hues and the more or less curved lines on their wings mimic dead leaves. A band of veins that is wider and perpendicular to the others, like the midrib of a leaf, heightens the impersonation. The *Cimicodes* belong to the vast family of geometer moths, which are small moths numbering more than 20,000 species worldwide.

Following spread

Here we see "the most wonderful case of defensive mimicry in a butterfly," according to the naturalist Alfred Russel Wallace! The genus *Kallima* (here, *Kallima paralekta*) includes about 10 species originating from Asia; the shape and colors of its wings, once closed, have the exact appearance of a dead leaf. Even within the same species, each butterfly has its own cryptic patterns that are unique to them, so predators do not become accustomed to a single morphotype.

MOVING TWIGS

Cryptic plumage occurs in many birds, particularly females that brood on the ground. Few birds, however, are able to pretend to be what they are not. The potoo is one such bird. Nocturnal hunters with big yellow eyes, these close relatives of nighthawks share mimetic coloring with the latter, but some also practice homotypy. Thus the gray potoo can often be found perching at the end of a dead branch during the day. It will then freeze at the slightest hint of danger, its head slightly raised and its eyes closed or cracked open just enough to observe the surroundings. Perched a dozen or so feet from the ground in the light filtered by the foliage of the trees, it remains incognito, looking like a dead branch. It is so sure of its camouflage that it does not bother to hide itself to brood. Instead, it lays its single egg in the slight depression of a branch, and both parents take turns sitting on it in an almost vertical position, a behavior that the chick adopts as soon as it is alone.

Leaves, branches, twigs and other bits of wood are among the most often imitated plant models, and once again it is insects that are the most skilled at this form of deception. Such homotypy is particularly common in lepidopterans, whether at the caterpillar stage or the butterfly stage. Thus, when it is at rest, with its wings along the length of its body, the buff-tip moth looks like a piece of dry twig. Moreover, the caterpillars of the geometer moths, which belong to a very large family of nocturnal moths, can take on the appearance of a small branch while standing erect and motionless for hours, like twigs among twigs. Some mantises are also masters of the art of getting confused with twigs. This is the case of the African mantis *Popa spurca*, whose camouflage is intended to deceive both its prey and its predators.

But the twig insect par excellence is, of course, the aptly-named stick insects. Of the nearly 3,000 species listed to date, about 40 (in the *Phyllium* order, see pages 59 and 61) are leaf imitators, while all others mimic twigs to perfection (these are sometimes called "walking sticks"). This is an art that the Phasmida have been cultivating for over 120 million years. Three specimens from a fossil species called *Cretophasmomima melanogramma* were found in China in a deposit from the late Cretaceous period. In 2014, the researchers demonstrated that this species had a leaf-like pattern similar to the deposit itself, with similar dark bands. This discovery shows that the imitation of plants began very early in stick insects, probably under the pressure of predation by birds and insectivorous mammals, whose diversification during that period is well documented.

Everything about stick insects is reminiscent of twigs: their slim, tubular bodies, their slender legs and the appearance of their epidermis, which in some species is covered with thorns and can vary in color according to the season. In *Clonopsis gallica*, the color tones are consistent with those of the surrounding vegetation, which always tends to darken during the summer. Despite the perfection of their camouflage, stick insects remain cautious. Most species are nocturnal, spending the day hidden under bushy cover.

Why so much prudence? Because predators are not so easily fooled. Experiments conducted in the 2010s at the University of Glasgow by John Skelhorn, Graeme Ruxton and their colleagues brought together chicks, twig-mimicking caterpillars and twigs. The chicks who had been shown real twigs showed no interest in the twig-imitating caterpillars, having found nothing of interest in the real thing. The subterfuge had worked. On the other hand, the chicks that had never seen twigs devoured the caterpillars... The experimenters also showed that if the caterpillars were presented together with the copied twigs, the

With a height of nearly 24 inches (60 cm) and a wingspan of 29½ inches (73 cm) the great potoo is the largest of the potoos (nocturnal birds of Central and South America). Its size does not prevent it from going completely unnoticed during the day; when it stands still, usually in a large tree, its cryptic plumage works wonders. It is only at dusk that it takes off to hunt, with its large beak open like a nighthawk, gobbling large insects in flight.

chicks learned to detect the difference. In other words, if the model and its imitation are together in one place, a predator can compare them and identify the prey. We now better understand the caution of the stick insect.

As much as the copy looks real to us, it is never infallible, and camouflage is only one means prey have at their disposal to defend themselves against their predators. The few species of stick insects that have wings, which are often colored and spotted, will open them suddenly to scare predators. Moreover, all stick insects practice autotomy, voluntarily abandoning one of their legs to their predator. Finally, when disturbed, they are able to feign death (called the thanatosis phenomenon): They fall down and are completely immobilized, entering a state of catalepsy.

One would swear at first sight that this is a very small, bevel-cut piece of wood. This perfect copy leads one to simply ignore the object. Present throughout Eurasia, the buff-tip moth frequents a wide variety of environments, such as forests, parks and fallow fields. It is sometimes called the bull's head moth, since it carries its head down, like a bull ready for battle.

This "snapped branch" is the chrysalis of one of the largest North American butterflies, the giant swallowtail (*Papilio cresphontes*, not to be confused with the similarly named Old World swallowtail). It spends its winters in the form shown here, stitched to its branch with silken threads. When warm weather returns, the imago emerges and reveals its black and yellow wings, with a wingspan that can reach up to 5 inches (12 cm). It is not only at the chrysalis stage that this species pretends to be what it is not. As a caterpillar, it looks like bird droppings and gives off a nauseating odor when disturbed.

Above

And what if the trickery were not only visual but also olfactory, or perhaps involved taste? The caterpillar of the giant geometer moth (*Biston robustum*) absorbs the chemical components of the plants on which it sits. It then stores these components in its cuticle, which takes on their taste and smell. This chemical disguise allows the moth to go undetected, even by predatory insects, such as ants, which will run across its back as if it were just a twig.

At left

Motionless, stiff as a board, the caterpillar of the swallow-tailed moth is the perfect imitation of a twig. It anchors itself to branches and twigs with its false legs, located at the back of its body, which form a hook. Its true legs are at the front of its body (in the photo at left, they are bent against its body). It can stay like this all day, totally rigid, waiting for the nighttime hours to feed. This moth also spends the winter in caterpillar form, but the larva takes care to hide itself in a cavity within the bark of its host plant.

Above

The African twig mantis (*Popa spurca*) adopts the same camouflage tactics as stick insects. However, while stick insects use these tricks to avoid being detected by predators, the mantis also uses mimicry to surprise its prey. It can remain motionless for long periods of time, and when it needs to move, it does so slowly. This mantis is up to 3 inches (8 cm) long, with males being about ½ inch (1 cm) smaller than females.

At right

At 4 inches (11 cm) long, *Megaphasma denticrus* is the largest of the North American stick insects. It is also the most colorful: golden brown stained with green and some red marks. This coloring does not prevent it from going completely unnoticed in the middle of dry grasslands, however. Males are rare in the wild, where a ratio of one male per thousand females seems to be the rule. This does not prevent the females from laying between 100 and 150 eggs during their lifetime.

Grasshoppers of the genus *Paraphidnia* are easily recognizable. Their elongated, narrow wings slope at a 45-degree angle to their abdomen, which gives them a twig-like aspect. The colors and textures of the specimen shown here also allow it to blend in perfectly with the mosses and lichens on which it has settled. These insects fly well and live in the canopy of tropical forests, mainly in Central and South America. New speciesof this still poorly known genus are being discovered regularly.

DISAPPEARING IN THE DEPTHS

If a prize for animal deception were to be awarded, the Antennariidae (anglerfish and frogfish) would get the top prize. These coral reef fish use the full range of impostor tricks to perfection. They use disruptive colors, ocelli and other ploys to deceive their opponents and to attract their victims. They also display a much wider range of heterogeneity than most other species. Simply put, anglerfish and frogfish provide a primer on camouflage all by themselves. The stocky body of these poor swimmers looks like a pebble covered with seaweed or some harmless invertebrate like a tunicate, coral or sponge.

At the bottom of the sea, in the heart of the reefs, like on the Earth's surface, a struggle for life plays out. And in this incessant fight, some seem quite helpless. The pink pygmy seahorse is ¾ to 1 inch (2 to 3 cm) long. Lacking protective scales, claws, teeth and venom, it has no *apparent* chance of survival in this world, where the biggest eat the smallest! Its saving grace is that its existence is limited to the contours of the gorgonian branches of corals of the genus *Muricella* and that it copies its surroundings perfectly thanks to its shape and colors, including the tubes adorning its body, thus becoming invisible. This mimicry is reminiscent of that of the candy crab, whose spiny body with mostly white and fuchsia pink hues faithfully reproduces the shapes and colors of the soft coral that shelters it.

Much like on solid ground, marine imitators often favor plants for their models. Close relative of the seahorses, the leafy sea dragon owes its name to the long, leafy outgrowths it bears on its whole body. With its seaweed look, this fish, which can exceed 12 inches (30 cm) in length, lives in the vegetation-rich waters along Australia's southern coast. Ghost pipefish, tropical sea fish that are close relatives of seahorses, are not to be outdone when it comes to camouflage tricks. The robust ghost pipefish has a thin and laterally compressed body that varies in color (green, brown, orange, black,

etc.) and is sometimes covered with small spots reminiscent of parasites that attach themselves to leaves, giving it the appearance of plant debris floating in the water.

Hippolyte inermis which can be found in European seas, displays a fairly similar form of homotypy. This small shrimp is usually green, sometimes with pink spots that imitate the red algae that frequently lives on plants in those waters. Standing motionless, it looks very much like a little bit of eelgrass or posidonia, which are among the few flower-bearing marine plants. Its coloring changes based on the time of day and the season. In winter, brown, yellow, purple and black specimens are more common. But unlike other shrimps, it cannot take on any color at will. Each one must choose a surface that matches its coloring. It is hypothesized that it perceives colors through its pedunculated eyes, although some doubt remains.

Even greater is the mystery of the emerald elysia and the green elysia. These sea slugs have the green color and overall appearance of a leaf. However, does this resemblance help them hide, or is it the result of evolution? In fact, this mollusk feeds on algae, from which it collects chloroplasts, the cells that enable photosynthesis in plants, which it stores. For 40 years, researchers have been trying to understand how elysias practice photosynthesis, like the leaves they have come to resemble. But no matter how these mollusks feed on sunlight, their homotypy has earned them the nickname "small crawling leaves."

Everything must disappear: This is the principle of the camouflage adopted by the warty frogfish pictured here. While it is stationed rather prominently on its rock, it is still quite difficult to locate. The tints and textures of its body beautifully match the substrate.

Previous spread
Stationed on a rock, this De Beaufort's flathead, also known as the crocodilefish, waits patiently. Its flat head and body, as well as its large pectoral fins, obscure its fish morphology. But its coloring, composed of brown and greenish spots intertwined with lighter curved lines, gives it perfect camouflage. This Indo-Pacific species, which reaches a maximum of 20 inches (50 cm), feeds on fish and crustaceans.

It is at the bottom of the oceans that we find the most incredible cases of homochromy with non-uniform color tones. Some species of fish, like the California scorpionfish shown here, seem to have a limitless ability to adopt the most varied mixtures of colors to make themselves as invisible as possible.

Totally indistinguishable on its rock, the red lionfish waits patiently. It is waiting for a small fish, crustacean or mollusk to come within reach of its protractile mouth. It will only take a few tenths of a second for the fish to make a meal of it. It has so much confidence in its camouflage that it is easily approached, but a word of warning to any diver who might put their hand on it: The spines that cover most of its body contain venom that injects itself by simple pressure.

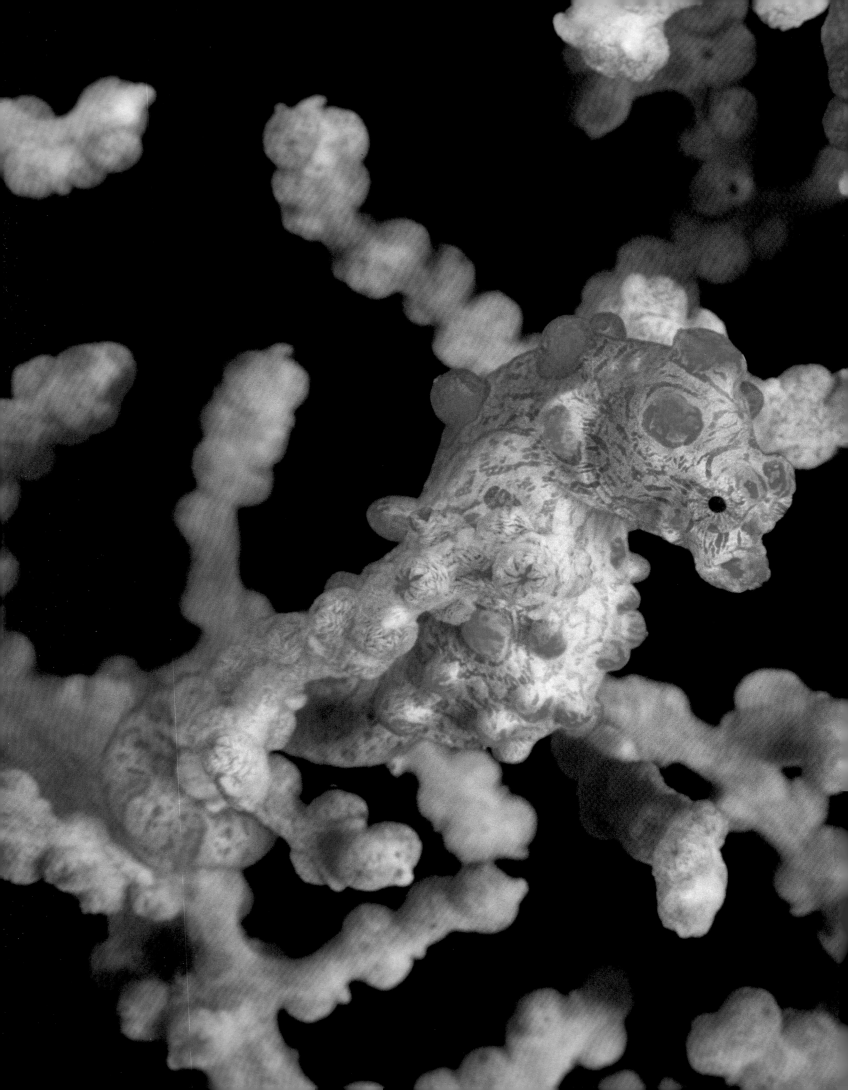

At left
The pygmy seahorse, also known as the Bargibant's seahorse, never strays far from its branch of sea fan, clinging to it as if to a lifeline. In any event, its atrophied fins do not allow it to make large movements. However, this does not prevent it from hunting the zooplankton that passes within its range. The seahorse relies on its perfect mimicry to avoid being detected by potential predators. The bulbous growths adorning its body perfectly copy, in shape and color, those of the sea fans of the genus *Muricella*.

Following spread
This is undoubtedly one of the most difficult crabs to photograph. *Hoplophrys oatesi* — the candy crab — measures 1 inch (2 cm). Founc from the Red Sea to the Maldives and across the Pacific, this crab frequents coral reefs, where it takes refuge in the heart of soft corals (like *Stereonephthya*, pictured on the following spread), its perfect camouflage giving it superb protection.

You cannot confuse the leafy sea dragon with any other fish. Its appearance is unique in the marine world. A close relative of seahorses, this inhabitant of the Australian coast is, however, much larger, and it easily reaches about 12 inches (30 cm) in length. But what makes it truly special are the long protuberances that adorn its body and the leafy appearance that helps keep it concealed in the algae beds where it takes refuge.

Body and fin shape, green color, texture, vertical positioning...
Everything is there. The halimeda ghost pipefish (*Solenostomus
halimeda*) has a very specific mimicry model. This small, 3-inch
(7 cm) fish resembles only one kind of seaweed, with which it
shares its name: *Halimeda*. The ghost pipefish generally positions
itself upside down and sways along with the currents. It is found
in coral reefs in the Indian and Pacific Oceans.

Another ghost pipefish, the rough-snout ghost pipefish, also copies plants. It can vary in color (yellow, green, black or brown), and its flat and elongated body, 7 inches (17 cm) long, looks like a piece of aquatic plant set adrift. The illusion is reinforced by the presence of spots mimicking the presence of parasites. Often, it looks like two pieces of seaweed gliding side by side. It seems that ghost pipefish are monogamous; they only live for one year, dying after a single season of love...

Above and at left
A close relative of the crinoid shrimp (see page 21), this cleaner shrimp of the genus *Laomenes* does not leave its crinoid (which is an animal similar to sea urchins and starfish). While many marine species live symbiotically with cnidaria or echinoderms, it is the shrimp that most frequently lives this community lifestyle. In terms of camouflage, should we talk about mutualism (both organisms benefit) or commensalism (only the shrimp gains an advantage, but it does so without disturbing its host)? Scientists are still trying to explain the terms of the contract.

This Xenia shrimp (*Hippolyte commensalis*) is perfectly concealed in its soft coral of the genus *Xenia*. Until the 2010s, only one species of this genus was known to have a commensal relationship with soft corals. More recent research has revealed that no less than three species share this type of relationship with soft corals. Scientists believe that the frequency of commensal relationships between animals and coral reefs contributes to the biological diversity of this biotope. This biodiversity is almost certainly underestimated, and there is still much to learn about it.

Above
With an average size of 1 inch (3 cm), green elysia (pictured above) and emerald elysia look like small lettuce leaves. But this is not all that they have in common with vegetation. These sea slugs, like plants, are capable of photosynthesis, which allows them to enrich their menu with carbohydrates. Both live in shallow waters on both sides of the Atlantic.

At right
The diamond leatherjacket (*Rudarius excelsus*) does not exceed 1 inch (2.5 cm) in length. It looks very much like small round leaves, since only its body is visible — its caudal and dorsal fins being transparent — and it can pass completely unnoticed amidst the intensely green oval leaves of *Halophila ovalis*. Found in the western Pacific, from the Sea of Japan to the Great Barrier Reef, it inhabits shallow waters, where it feeds on plankton.

DECEPTIVE IMAGES

What is the best way to avoid predators? Simply by being an inedible, inert object without any nutritional value. In areas rich in vegetation, grasshoppers and locusts mimic the leaves around them to perfection. However, in the rugged deserts of Southern Africa, they have to change their strategy. Thus, the grasshopper *Trachypetrella anderssonii* looks like a small stone in the vast desert. Its squat, angular body makes it totally indistinguishable from its stony environment; only movement can betray its presence. In the bottom of the sea, there are other fake pebbles. For example, the crabs from the Parthenopidae family, such as *Daldorfia horrida,* have a shell with a grainy, furrowed, worn appearance, which looks like a small rock in the midst of the reefs and the aquatic plants that colonize them. Another member of this crab family, *Cryptopodia fornicata*, which lives on the sandy bottoms of different oceans, also has a mineral-like appearance. Its smooth, domed carapace, which completely conceals its legs and against which it folds its powerful pincers, looks like a rock. More unexpectedly, crabs of the genus *Oreophorus* have a shell carved with furrows and alveoli, imitating the dead corals among which they live.

One of the most original and certainly most unexpected homotypes is undoubtedly taking on the appearance of bird droppings. A small, fresh dropping left on a leaf by a passing sparrow? No, the caterpillar of an Asian butterfly, *Papilio lowi*. Everything is there: the color, the shape and even the smell. When it is attacked, the caterpillar emits a nauseating odor that is not at all appetizing... This imitation is found in several species of Lepidoptera. In large caterpillars, this form of mimicry is only encountered during the animal's early stages,

Pebbles among pebbles — the grasshoppers of the Pamphagidae family, found in arid areas of Southern Africa, go completely unnoticed, so confusing is their mimicry of stones. It is usually only when they jump that they reveal their true identity. In these vast plains where vegetation is scarce and wind is omnipresent, they have lost the ability to fly, their wings having become vestigial.

when the larva is not yet very active. The effect would hardly be believable in a large larva bustling to find its site for nymphosis! Some adult butterflies, including species from the United States, such as *Eudryas unio* and *Ponometia elegantula*, once settled, also take on the appearance of excrement. Among the vertebrates, a small African frog, *Afrixalus pygmaeus*, does not have any objection to being mistaken for poop!

To test the effectiveness of this disguise, I-Min Tso's team at Taiwan's National Chung-Hsin University studied the *Cyclosa ginnaga* spider. This spider must attract its prey to its web without being noticed by its main predator, wasps. The arachnid weaves a white disk and

positions itself in the center, the whole scene thereby resembling bird droppings, at least from a human's point of view. But how do we know whether the spider's predators see the same thing? The researchers compared 125 disks with 27 bird droppings and found no statistical difference in terms of their size. In the laboratory, they conducted different experiments, removing the false droppings by dyeing them black and leaving their host visible or not. When the spider was easily identifiable, it was consistently attacked. On the other hand, the wasps lost interest in the webs that were left intact, no doubt assuming "this is not a spider" — a deceptive image indeed.

Above
Examples of collective camouflage are rare. One of the most
famous is that of the treehoppers (genus *Umbonia*). Grouped
around the same stem to feed on the juices of the plant, these
insects take on the appearance of thorns, which cannot, under
any circumstances, be of interest to an insectivore! But for this
type of homotypy to work, the insects must live in groups,
since a single thorn on a branch would be suspect.

At left
Perfectly homochromatic in its stony universe, this other
grasshopper of the Pamphagidae family has pushed mimicry
so far its body is as angular as the small stones among which
it lives. At the slightest danger, it folds its antennae against its
body and remains perfectly motionless. This grasshopper lives
in semi-arid areas of Southern Africa.

Above

No, this is not a deposit from a careless bird. This small "bird dropping" is actually a caterpillar of the alder moth, which, as its name suggests, uses mostly alder trees as host plants. The appearance of this caterpillar is unappetizing, but it offers it great protection. The caterpillar only uses this camouflage in the early stages of its development. As it grows, it becomes black streaked with yellow, and its body is adorned with long black hairs.

At right

A resemblance to excrement is useful to both avoid predators and attract prey. No insectivore is interested in bird droppings. On the other hand, several insects can find nutritional value from it — and they pay for it with their lives. The spider *Phrynarachne decipiens* has the colors and sheen of freshly deposited guano. To perfect its deception, it weaves a creamy and asymmetrical white web around its body. The illusion is all the more perfect as it gives off the same smell as its model...

A FOOL'S GAME

Homochromy and homotypy are not the only means by which animals camouflage themselves. They also know how to bluff, lie and disguise themselves in order to destabilize their prey and predators. Ocelli, which are big round spots that look like eyes, are one of the most common theatrical techniques! But how are they used by the prey? How are they perceived by the predator? How effective are they? These questions are now the subject of many studies.

Animals are also actors, able to play the role of something else, dressing up like corpses, dust or twigs. Some even physically transform themselves to look like another species. Many insects have the ability to mimic ants, both physically and behaviorally. This phenomenon is so widespread in nature that it has a name: myrmecomorphy (from the Greek *murmex*, which means "ant"). And what about a caterpillar posing as a snake? It's a fool's game — where the ability to fool can save one's life!

Previous spread
"Let's go out under cover" could be the motto of the crabs from the Dromiidae family. Like the *Austrodromidia octodentata* pictures on the previous spread, these crabs are usually covered with a sponge. This hides not only their body but also their odor.

At right
Take a good look at this picture! Do you think you see an ant? Count the number of legs — there are eight! This is a spider. It belongs to the large Salticidae family, which are species known for their ability to jump. This arachnid's entire body has been transformed to maintain the deception. The most visible change is the constriction of the abdomen so that it is the size of an ant's. To increase their likeness to ants, most myrmecomorphic spiders adopt behavior similar to that of their model.

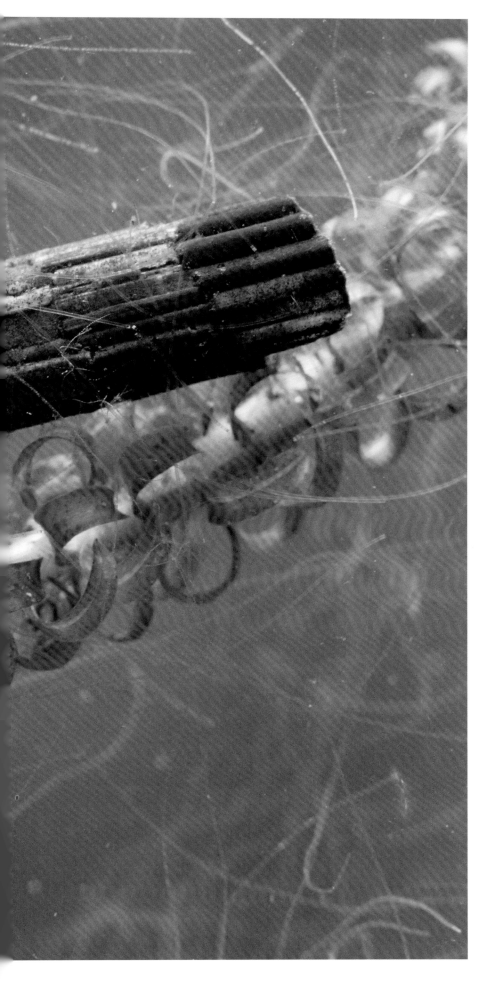

AT THE MASKED BALL

"In its natural pond, the caddisfly does not often have the choice bits of wood that I have offered it in the drinking glass. It can find a little bit of everything and makes use of all of it just as it is. Pieces of wood, large seeds, empty seashells, bits of straw, and shapeless fragments take their place in the construction, for better or worse, without touching the saw; and from this amalgam, the fruit of chance, results an edifice of shocking error."

The man who so elegantly described the work of the caddisfly is the famous entomologist Jean-Henri Fabre (1823–1915). The caddisfly and its work is one of the many insects he described with equal parts precision and talent. Caddisflies are small insects that look like moths but constitute another order altogether: Trichoptera. They are also, like moths, irresistibly attracted to artificial lights. For the moment, it is not the adults that interest us, but the larvae, which live in fresh water. Fishermen, who use them as bait, have dubbed them "log trays," "wood holders," and "leaf holders." It was the naturalist Pierre Belon (1517–1564) who named these larvae after the Greek word *phrúganon*, which means "small dry wood" and became a name to designate Trichoptera in general.

Why these names? And why do the larvae of these insects arouse such interest? Because of their carpentry. Indeed, in most species of caddisfly, the larva builds a protective case that allows it to pass unnoticed by fish and other aquatic predators. Thanks to special glands, called silk glands (which produce silk), it weaves a "shirt" that it then covers with little bits of whatever it finds in its environment. It uses its silk to affix these bits and pieces, creating a protective cover for itself. Some

What is probably most surprising at first sight is the uniformity of the construction. Each small piece of stem is precisely placed so the case forms a near-perfect tube. Even more impressive is the fact that the caddisfly larva must build a case that is light enough to be transported, solid enough to withstand the current and have just to right amount of buoyancy, being neither an anchor that sinks to the bottom nor a buoy that floats. Everything is maintained thanks to silk threads capable of sticking under water, which is a real technical feat, since water and glue do not usually mix well...

species use only minerals, while others prefer vegetable matter. A third category does not care about the type of material as long as it can get its whole abdomen in the tube. The caddisfly hold its case in place with two claws at the end of its soft, otherwise unprotected body. The larva, however, is not permanently attached to its tube; it can get rid of it to make a quick escape. It then goes back to work, building a new case, which it abandons completely when it becomes an adult. Caddisflies were once so numerous that their fossilized tubes have formed

geological layers called "caddisfly limestone," which are dated to 25 million years ago.

Caddisfly larvae are not the only larvae to form a protective case using material from their immediate environment. This tactic is also found in bagworm moth caterpillars. In the larval state, these little moths live in a silk case made of materials from their surroundings. Each species has its favorite material: sand, small pebbles, leaves, lichens, tiny gastropod shells. The cases and their usage can also vary by sex; while the males get rid of it

upon maturity, the females do not. Wingless and vermiform in their adult state, the females spend their whole life inside their case, even laying their eggs inside it.

Many other species use material from their environment to carve out a made-to-measure suit and camouflage themselves. Among them is the masked hunter, a small bug that is also known as the bed bug hunter and has a strictly carnivorous diet, unlike most of its congeners, which are vegetarian. Equipped with a rostrum, it "stabs" its victims and sucks out their organs. This method of execution gave

the family its common name: assassin bugs. Larvae and adults use this same feeding method. However, while the adult goes out uncovered, the larva is camouflaged. It can cover its body with dust thanks to its integument, which is coated with a sticky secretion that, with the addition of the dust, creates an inconspicuous covering.

Some spiders also hide their body with the soil around them. In species of the genus *Cryptothele*, microscopic, curved, barbed hairs keep the particles on the arachnid's body. In other species, an adhesive secretion may also be involved. This seems to be the case for a Mexican species first described in 2014, *Paratropis tuxtlensis*, which is covered with particles of earth embedded in its cuticle. Adhesion is also apparently provided by a glue it secretes. When alarmed, it remains motionless, relying on its perfect camouflage for protection.

Some animals are more selective in their choice of costume. One of the most peculiar, some might even say morbid, of these disguises leads us back to bugs, this time to an assassin bug from Malaysia, *Acanthaspis petax*. Like its masked counterpart, it consumes the

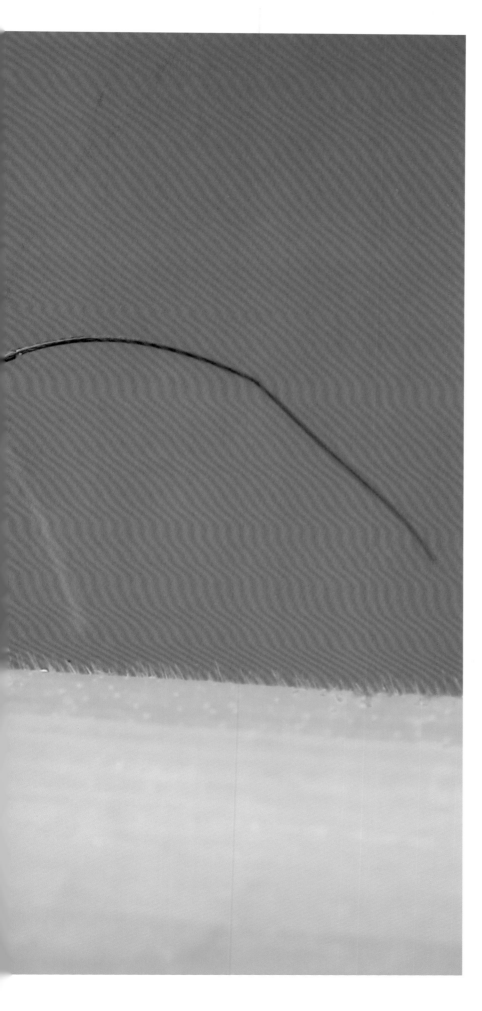

liquefied entrails of its victims, but instead of abandoning the empty envelopes, it sticks them on its back and can lug around 20 or so heaped remains! With ants being among this bug's prey of choice, it is not impossible for this camouflage to attract more of its victims, which, recognizing the smell of their congeners, may come close and in turn become the slayer's prey.

Self-adorning with one's emptied victims is not just the privilege of assassin bugs. At the bottom of the oceans, mollusks of the genus *Xenophora* cement empty shells that they find on their travels onto their own shells. Some species simply adorn the edge of their shell to break their obvious shape, while others make full decorations that conceal them completely.

Let's stay at the bottom of the ocean, since there are some masters of the masquerade to be found there. *Dromia* are a genus of small, round, hairy, stocky crabs, many of which use camouflage. The sponge crab (also called the sleepy crab) takes a sponge, cuts it to fit its size and holds it to its shell using its two back legs, which are hooked and curved upward. The sponge, which could have spent its entire life in the same spot, doesn't have it so bad, as it gets to wander along with the crab. As the sponge grows, it will eventually conform to the exact shape of its host. Not only does this hide the crab from predators, it also keeps them away, since few enjoy the taste of sponges.

In spider crabs (genus *Maja*), many species conceal their shape under accumulations of algae, pieces of sponge, sea anemones, and the like, which they affix to their carapace with hooked bristles that act like Velcro strips. Some species have perfected the subterfuge by

After having liquefied and swallowed the entrails of its victim, the larva of this assassin bug, *Acanthaspis petax*, has accumulated the empty cuticles of the ants on its back. According to a study conducted by New Zealand researchers, this macabre display appears to prevent attacks from the larvae's worst enemies, the spiders of the Salticidae family. The latter hunt by sight, so they are unlikely to identify this pile of corpses as potential prey.

It is very difficult to see the crustacean in this photo! Crabs of the genus *Huenia* disguise themselves as an aquatic plant. Not only is their green body covered with seaweed, but some perfect the masquerade by affixing small round leaves of plants from the *Halimeda* genus at the end of their pointed rostrum. These crabs are found in the reefs of the Indo-Pacific basin. With a carapace measuring ¾ inch (2 cm), they can go completely unnoticed. Some species close to *Huenia* prefer hydrozoa over algae.

adding a chemical shield; the decorator spider crab is adorned with poisonous corals, while *Libinia dubia*, another spider crab, wears toxic algae to keep fish and octopuses at bay.

Other animals live symbiotically with algae in order to remain as inconspicuous as possible, but it is necessary to rise to the heights of the treetops in the American tropics to encounter the sloth, the mammal known for its extreme slowness. Among the many small organisms that live in its coat, which is a true ecosystem of its own, we frequently find *Trichophilus welckeri*, a green alga that colors the sloth's fur, allowing it to blend with its surroundings as it spends up to 20 hours a day sleeping on a branch.

Top right
By adorning their shells with sponges, tunicates, anemones and soft corals, the crabs of the Dromiidae family ensure they are effectively protected, as they benefit from these organisms' venom. Such relationships can be described as mutualism, since the sponges and other passengers benefit from the travels of their "vehicle," which helps them diversify their food sources and more widely disperse their gametes.

Bottom right
Concealed from predators and confident in their camouflage, crustaceans that adorn their carapace, such as this decorator crab, appear to be more diurnal than most other crustaceans.

Following spread
The decorator crab, which belongs to the vast family of spider crabs, covers its body with a little bit of everything that it finds, adorning its carapace with a real patchwork of shapes and colors. In addition to providing camouflage, this disguise, from which only the animal's pedunculated eyes stand out, also acts as a second carapace. This small crustacean measures 4 inches (10 cm), legs included.

IN THE EYES

Among the many patterns found in nature, ocelli have perhaps intrigued zoologists the most. These spots, which are more or less round and can be plain or gaudy, are always eye-catching. They are found in birds, fish and amphibians, but it is in insects, particularly butterflies, that they were first studied. The resemblance of certain ocelli to eyes did not escape biologists' notice, and the first studies on their possible role in prey-predator interactions were carried out in the 1950s by the English researcher David Blest. He used a fairly common European butterfly, the European peacock, whose wings have a cryptic brown underside and a brightly colored upper that is adorned with two pairs of ocelli. When attacked, it opens its wings quite quickly, revealing the ocelli and creating a hissing sound. The concept of the experiment was quite simple. The researcher placed peacock butterflies, some of which had their ocelli erased and others that were left intact, in the company of chickadees. He quickly discovered that individuals without ocelli were more often predated on by the birds as compared to the butterflies with their ocelli.

One question remained: Do chickadees really perceive eyes, or are they just frightened by the sudden appearance of bright colors? In 2015, ethologist Sebastiano de Bona and his team from the University of Jyväskylä, in Finland, exposed chickadees to various images, including owl eyes and different types of ocelli. The experiment consisted of allowing the chickadees to attack flour worms next to a computer. At the time of the attack, either the eyes of an owl or ocelli more or less similar to eyes would appear. The birds' reaction to the eyes and to the ocelli that resembled eyes were quite similar, and the scientist concluded that the ocelli created the same effect as the staring eyes of the chickadee's foe.

Some caterpillars are not happy with just exhibiting a false gaze. They transform their body into the scariest animal, the one with the most mesmerizing stare... The snake! In species such as *Hemeroplanes ornatus* and *Hemeroplanes triptolemus*, large South American sphinx

moths (also known as hawk moths), the ocelli are found at the back of their body. These caterpillars are able to inflate this part of their body, essentially transforming it into a snake's head. The size looks off to us, but the copy is disconcertingly realistic, especially as the larva can add a slight sway. In order to understand the effectiveness of such a bluff, two Canadian researchers, Thomas John Hossie and Thomas N. Sherratt, created several examples of artificial caterpillars, some with ocelli and some without and/or some with inflation and some without. They then arranged these "caterpillars" in the wild and left them at the mercy of insectivorous birds. Their results demonstrated that false prey with ocelli or inflation were less frequently attacked than those with neither characteristic. On the other hand, prey that could simulate a snake's head (having both ocelli and inflation) did not seem to gain any additional benefit, as the attack rate was similar to that of prey with only one of these characteristics. What interest does this masquerade offer, then? Could it be an escalation of evolution, or is the explanation simply still escaping scientists? The mystery remains.

In coral reefs, the comet fish (also called the marine betta) is considered to be a more dangerous species than it actually is. When threatened, it sinks its head into a hole, leaving only the back of its body sticking out.

At right
It is hard not to see the imitation of a stare in these large round spots. Ocelli are found on many insects (here, *Argema mimosae*, a large African moth). They are also found on other animals, such as fish, reptiles, amphibians and even birds.

Following spread
The "eyes" of this nocturnal moth, of the *Automeris* genus, seem to stare at us intensely... Most scientists agree that ocelli help protect animals from attacks. However, ocelli's evolutionary mechanisms and the ways in which the predators they are supposed to scare perceive them are still very mysterious. Since the 2000s, several studies have been conducted to understand the hows and whys of ocelli.

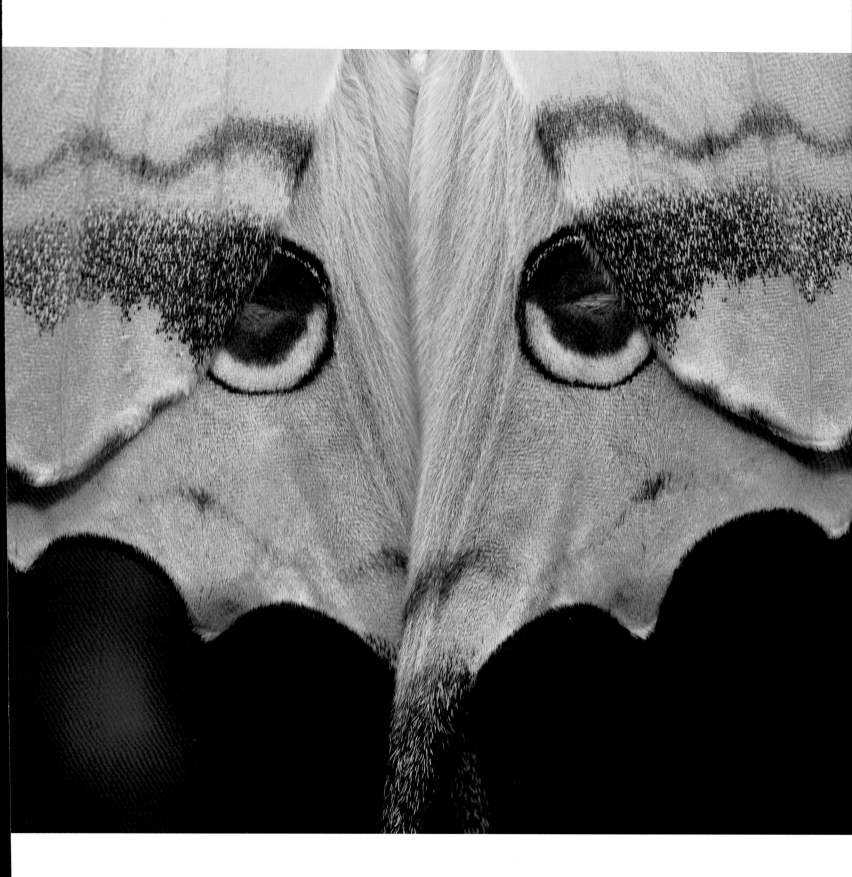

Thanks to its mottled skin, the way it positions its fins and the black ocellus serving as an "eye," the back of this fish resembles the head of the spotted moray eel. This beautifully speckled fish has a reputation for voracity, but when it doubt, it doesn't tempt fate.

Not all ocelli imitate eyes. Some are simple spots of color that bear no resemblance to any body part. In such cases, their function is to attract the attention of an attacking predator, which usually focuses on the most visible part of its prey. If these simple ocelli are located on the edge of a butterfly's wings or the tail of a fish, this is where the predator will strike first, giving its victim time to escape.

Who is this crab trying to scare with its red horned mask in the middle of a reef? The ocellate swimming crab belongs to a family of crabs that have legs that have become swimming paddles, which make them good swimmers. However, only the ocellate swimming crab has ocelli on its shell.

When threatened, the *Physalaemus nattereri* frog turns its back on its opponent and inflates its hindquarters, revealing a black-eyed mask. If this is not enough to push the enemy back, the eyes start to "cry" a kind of white secretion. The black ocelli ringed with yellow conceal what are actually toxic glands. This amphibian, which is about 2 inches (5 cm) long, lives in the humid plains of Brazil, Paraguay and Bolivia.

Some lepidopterans have refined their strategy. One notable case is *Remelana jangala*, an Asian butterfly with a blue upper wing and a rather dull brown underside, except at the extremities. These are not only decorated with bright colors but also with spots that resemble eyes, and they end in four thin black extensions that look like legs and antennae. The predator attacks this "false head," allowing the butterfly to take flight without risk of lethal injury. Many other species of the same family use this same ploy. Some, such as *Semanga superba deliciosa*, add a wing movement, which is perfect for focusing the predator's attention.

Above

The twin-spot goby is also known as the crab-eyed goby. Some have theorized that, when moving back and forth, it is imitating a crab. Given the number of species that predate on crustaceans, imitating one is perhaps not the safest way to avoid an attack. These ocelli would, however, almost certainly lead the enemy to a non-vital part of the body. They may also play a role during the breeding season.

At left

Imagine that you are faced with a harmless little dead leaf that suddenly turns into a monster with two large, staring eyes surrounded by bright colors. We don't doubt that you would recoil. This is the technique used by the peacock katydid to surprise its opponents. This South American orthopteran is up to 2½ inches (6 cm) long.

Above and at right

In terms of bluffing, some lepidopterans have set the bar very high. This is the case of the genera *Dynastor* and *Hemeroplanes*. Both include moths from South America whose imagoes are quite nondescript. The same cannot be said of the stages that precede them. When disturbed, the *Hemeroplanes* caterpillar (at right) inflates the front part of its body, which is adorned with two ocelli and resembles the shape of a snake head. You have to admit that the resemblance is striking. In *Dynastor darius* (above), it is the chrysalis, measuring about 1¹/₂ inches (4 cm), that looks like a snake's head, imitating even the large scales found on the head of real snakes (called cephalic scales).

TROJAN HORSES

A rather unusual moth was discovered in the Thai forests in 2005. *Siamusotima aranea* does not have ocelli-shaped patterns on its wings but rather "legs" surrounding a grayish mass. According to the researchers, these patterns mimic the appearance of crab spiders (Thomisidae family). Although this behavior has yet to be observed, it's theorized that this moth's appearance is designed to prevent attacks from real spiders.

This type of deception is well known in some moths of the genus *Brenthia*, such as *Brenthia hexaselena*. Pretty innocuous at first sight, this species stands all day on the surface of a leaf in an unusual posture for a lepidopteran: It keeps its four wings open and raised, forming a 45-degree angle with its body. If you look closely, the design on its fringed wings is reminiscent of the hairy legs of a spider. Certainly, to our human eyes, this imitation seems rough, but the butterfly is not content with such simple camouflage. On the leaf, it moves in small jerky jumps and abrupt changes of direction, which are not unlike those of jumping spiders (Salticidae family). These arachnids do not make webs but hunt by lying in wait. Their excellent eyesight allows them to spot their victim and assess the distance between them. Capable of jumping up to 40 times their own height, they throw themselves on their prey and then kill it with their venom. *Brenthia hexaselena*, however, has found a way to prevent such attacks: It pretends to be a predator. And to better convince its adversary, it never flees but instead goes to meet it as if it were a congener. And it works. The dupe in this case interacts with the false spider as if it were a member of its own species, going so far as to engage in a display to claim its hunting territory in the face of a rival it deems entirely too bold... Laboratory studies have shown that in the presence of a jumping spider, our plagiarist was not attacked, and therefore not devoured, in 93 percent of cases; other moths of a similar size that aren't imitators ended up as meals in 62 percent of cases!

While there is no doubt that some subterfuges are truly effective in misleading a predator, others leave the scientific world more perplexed. Such is the case for the *Macrocilix maia*, an Asian moth. A genuine image of two flies feeding on bird droppings appears on its outspread wings. This is at least the interpretation of some researchers; some writings even specify that the moth gives off a smell of dung. However, no study or observation in situ has yet confirmed this hypothesis. The truth might come from the tiny *Goniurellia tridens* fruit fly. Are the silhouettes of insects that seem to be drawn on its wings only the fruit of our imagination? A 2010 observation from the United Arab Emirates by Dr. Brigitte Howarth of Zayed University may well suggest that it is not. According to the researcher, as soon as the fly feels threatened, it waves its wings to indicate that it is not alone, taking advantage of the predator's confusion in order to slip away.

Above and at left
How can we tell whether this is simply chance or the fruit of our imagination? As pictured here, two flies appear to be drawn on the wings of the Asian moth *Macrocilix maia*. What's more, the two dipterans appear to be feeding on a bird dropping.
The mimicry of this moth, which does not copy an animal or vegetable species but a scene of ordinary life, is probably unique, especially since it is, according to some authors, accompanied by the proper smell. How does this imitation work? Does it really serve to distract a predator? Further research is needed to unravel the mystery of this moth with the realistically painted wings.

IDENTITY THIEVES

Some species of arthropod are masters of the art of taking on the physical appearance of another. The most copied models are ants, and this particular type of mimicry is called "myrmecomorphy." Among these imitators, we find our little jumping spiders (see page 140). In the large Salticidae family, the genus *Myrmarachne* comprises 200 species, all of which resemble ants in their morphology and color.

On the surface, this is not a simple phenomenon. Let's keep in mind that spiders have a body divided into two parts (a cephalothorax and an abdomen), while the bodies of ants and other insects have three distinct parts (the head, the thorax and the abdomen); ants have six legs, while spiders have eight; and finally, antennae play a vital role for ants, while spiders have none at all! But appearance is not everything, and most myrmecomorphic spiders move and behave like their model.

Myrmarachne melanotarsa, an African species, has taken the resemblance so far that it lives in small colonies of a few tens to hundreds of individuals, simulating life in an anthill, even though spiders are mostly solitary animals. Even more surprising, in some species of spider, males, females and juveniles do not imitate the same species of ants!

What benefits do spiders get from this perpetual disguise? It could allow them to approach, even to penetrate, an anthill without raising suspicion and, therefore, to consume ants. However, according to most scientists, this is instead a Batesian mimicry, in which a harmless species takes on the appearance of a species that is dangerous for its predators. Indeed, ants are rarely preyed upon. They are equipped with effective defenses, including toxic substances such as formic acid and powerful

Look for the intruder. Among these Asian weaver ants, hides a spider. The "ant" under the leaf does not have antennae and has eight legs, so it is indeed a spider, in this case a jumping spider of the genus *Myrmarachne*, and it is certainly a female, as the males have imposing chelicera that make them immediately identifiable.

pincers, and can respond in number to the attack of a predator. Laboratory studies by Ximena J. Nelson, of the University of Canterbury in New Zealand, have shown that *Portia fimbriata*, a jumping spider predatory to other jumping spiders, avoids ants as much as myrmecomorphic spiders.

Although there are an estimated 300 species of spiders disguised as ants, that is not exceptional. Nearly 40 insect families have species that more or less closely resemble ants. Beetles of the family Anthicidae are included in this number, but their copies are far from consistent and are

generally limited to the head and thorax. More surprising is the similarity of *Myrmecosepsis hystrix*, a wingless fly from Taiwan, to an ant. Some bugs in the Miridae family mimic ants, but only at the juvenile stage. These include *Himacerus mirmicoide*, which looks like a small

Here, the spider is second in line. It is holding its first pair of legs as if they were antennae. Jumping spiders of the genus *Myrmarachne*, which resemble weaver ants, do not live in the anthill but in the periphery. This proximity provides protection against predators, who fear the painful bites of weaver ants.

black ant. We also find this Batesian mimicry among the young of species of mantis of the genus *Acontista*. The juveniles, which do not exceed ⅛ inch (4 mm) at birth, copy ants even in their movements; they do not adopt the vertical position of adults but run horizontally and in a jerky fashion. They keep up their myrmecomorph until they molt for the last time. As adults, they take on every aspect of a real mantis!

Ant? Spider? Alien? Anyone who can clearly identify what the lobster moth larva looks like in a defensive position must be very clever. It is certainly the only caterpillar to have thin, very long, spider-like thoracic legs. Add to this the fact that its last abdominal segment is bulky, widened and flattened ventrally, that it has a sort of bifid appendage that looks like a hook, and that its head could be that of an ant. When threatened, this

Is it an ant? No, it's a bug of the genus *Hyalymenus* — more precisely, its nymph. In these insects, only the young look like ants. They can be found feeding on the same plants as their model, day and night, while the adults are nocturnal. They move their body and antennae similarly to their model. Depending on their stage of development, the nymphs copy different ant species.

harmless beast spreads and shakes its big legs, raises its abdomen and darts its hook about... In this hybrid composition, some see an ant, others a more arachnomorphic being. In any case, this frightening illusion, which measures 2 to 2½ inches (5–6 cm) seems to stop both birds and humans, who very often hesitate to touch it. Do we perhaps share certain ancestral fears of monsters, the deformed, the abnormal, that have been inscribed in the depths of our common history?

Above
In certain insects, myrmecomorphy only happens one time, in their youth. This is the case for several species of grasshopper in the Tettigoniid family. In the early stages of their life, juveniles copy the form and way of life of the ants among which they mingle. As an adult, the "little black ant" will become a grasshopper (maybe green) and, if it is a male, will chirp all summer.

At right
The bottom photo shows a young grasshopper of the genus *Polichne*. It only resembles its model, the weaver ant *Oecophylla smaragdina* (top photo), during its first juvenile stage, that is, when it comes out of its egg. From its first molt, it will begin to take on the morphology specific to Orthoptera, finishing in the form of a large grasshopper.

Above
At first, one hesitates a little regarding the identity of this creature. It is a bit like an ant, but not quite. The Membracidae belong to the order Hemiptera (aphids, cicadas, etc.). They are characterized by extravagantly colored and strangely shaped excrescences (outgrowths) on their thorax. Entomologists are not entirely sure what role these projections play. In some species of the genus *Cyphonia* (here, *Cyphonia clavata*), it seems obvious: The outgrowths resemble an ant, allowing the insect to protect itself from predators.

At right
The lobster moth's appearance with its plumed "tail" and russet coloring, has earned it the name *écureuil* (squirrel) in France, while it gets its English name from its resemblance to the crustacean. It is difficult, however, to specify which animal (or monster?) this harmless caterpillar hopes to look like when it puts itself in its defensive position.

LURES AND BAIT

"Every naturalist has his favorite example of an awe-inspiring adaptation. Mine is the 'fish' found in several species of the freshwater mussel *Lampsilis*. Like most clams, *Lampsilis* lives partly buried in bottom sediments, with its posterior end protruding. Riding atop the protruding end is a structure that looks for all the world like a little fish. It has a streamlined body, well-designed side flaps complete with a tail and even an eyespot. And, believe it or not, the flaps undulate with a rhythmic motion that imitates swimming."

The man who wrote these lines is Stephen Jay Gould (1941–2002). An eminent American biologist and paleontologist, he specialized in the theory of evolution, which he untiringly sought to popularize. Mussels of the genus *Lampsilis* were one of his favorite subjects of study and, especially, reflection. Unlike most bivalves, which entrust their eggs to the currents, freshwater mussels retain them until they hatch inside a pocket called a "marsupium." That's when things get complicated. Once hatched, the larvae, expelled from the marsupium, need to cling to the gills of a fish to grow. The *Lampsilis* thus "creates" a lure that looks like a small fish, a potential prey for many carnivorous fish. When such a predator presents itself, ready to bite into the lure, the mussel expels its offspring into the hungry fish's mouth...

What evolutionary process has allowed a mussel to create a fictitious fish? How did a totally blind mollusk with a rudimentary nervous system create a lure that mimics

Above and at right
When a single example of this viper was collected in Iran in 1968, herpetologists could not determine whether the strange appearance of the tip of its tail was a disease or a genetic abnormality. It was not until the 2000s that a new specimen was discovered, and it was understood that it was a species in its own right, which was named *Pseudocerastes urarachnoides*. The use of the end of its tail as a decoy was confirmed with animals bred in captivity before being filmed in the wild.

At left
What do you see in the photo? Probably a simple little fish. The most attentive of us can also perceive the presence of a mussel, whose tubes we can make out. Is the wavy-rayed lampmussel preparing to devour the unknowing fish? Not at all. This is a fake, a decoy created by the mussel. It is, in fact, an extension of its skin and is intended to attract carnivorous fish. The mussel uses this subterfuge not to satisfy its appetite, but to provide a home for its larvae, called "glochidia," which grow into juvenile mussels within the gills of fish. The juveniles then detach from their host and fall to the seabed and live independently. This amazing strategy is common to all mussels of the genus *Lampsilis*.

a fish credibly enough to fool predators? According to Gould, a part of the marsupium and mantle (the fleshy inner face of the shell) has evolved to give birth to the decoy. Evolutionists speak of "pre-adaptation": an organ existing in a primary form and for a given function that evolves toward an entirely different utility. This evolution is done in stages, without a predetermined goal. Here is an example of a step: Some species of mussels similar to *Lampsilis* have no decoy. On the other hand, they frequently agitate their marsupium, which agitates it and oxygenates their eggs. These muscles also allow the movements that the *Lampsilis* use to move their decoy.

But what pre-adaptation may have given birth to the spider-shaped decoy that ends the tail of the Iranian viper *Pseudocerastes urarachnoides*? Described for the first time in 2006, this champion of camouflage is easily hidden among rocks. Only the decoy, which it waves, remains properly visible. Its function is to attract the reptile's prey, which is insectivorous birds.

The alligator turtle (*Macrochelys temminckii*) uses its decoy in a similar fashion. Despite its size (it weighs up to 220 pounds [100 kg] and can be 30 inches [75 cm] long), it remains invisible in the bottom of the swamps of North America. Perfectly concealed by a homochromy reinforced by the presence of algae on its shell, it keeps its powerful mouth open. On its tongue is a decoy, clearly visible and red, that it waves like a worm. Rare are the fish, amphibians, turtles or even young crocodiles who, cheated, come out unscathed from the terrible vise...

In the same way, the trap set by Lophiiformes is just as destructive. Commonly called anglerfish, the species in this order are all poor swimmers and specialize in decoy hunting. For this purpose, the first ray of their dorsal fin has been transformed into a kind of rod ending with a bait called "esca" or "illicium." This order also includes the frogfish, which blend so seamlessly into their environment and with which we have already become acquainted (see pages 23 and 88). These anglerfish draw the attention of their future victims through their esca, which, depending on the species, looks like a worm, a shrimp or a small fish. They are not very fussy about their diet, taking only 6 milliseconds to devour their prey; they are the fastest gobblers in the ocean! The other big mouths in the sea of the same order, the monkfish of the genus *Lophius*, deceive their victims thanks to their filaments, formed from the first elongated ray of the dorsal fin, which they shake above their frightfully toothy mouths.

Close relatives, the Melanocetidae anglerfish have something more. In the underworld of the oceans, more than 1640 feet (500 m) deep, their bait shines in the night. Called a "photophore," this luminous appendage shelters millions of bacteria. Living in symbiosis with the anglerfish, these generate bioluminescence, a cold light that is present in 90 percent of deep-sea organisms. In return, the anglerfish provide the bacteria (which can live independently) the safety of a shelter and sugars and amino acids to feed them. While most decoys have been developed to trap victims, that of the anglerfish has a twofold function. It is equally likely to attract prey and males in search of sexual partners. Indeed, only females are equipped with a photophore, and to avoid confusion, each species has a light of its own. Once a male, which is much smaller, has found his female, he docks himself to the female by biting her body. Some of his organs then merge with those of his partner. He will now live as a kind of parasite for the sole purpose of providing sperm when the female is ready to spawn. Since that can take years, it's best to be there at the right time. This, in the immensity of the oceanic depths, remains rather random...

Bulging eyes, huge U-shaped lips turned upside down, a mouth filled with sharp teeth... Almost buried in the sand, the whitemargin stargazer is patient, waving its fleshy decoy. It will only take a few tenths of seconds to swallow a careless fish. The family Uranoscopidae can be found in all oceans and in many seas. There are about 50 species, some of which can discharge electric shocks to stun their prey before swallowing them.

This fish is nearing its end. In a few seconds a hopeless trap will close on it, but it only sees the little red worm moving... The alligator-snapping turtle takes its time. With its metabolism slowing, it can go about 50 minutes without breathing, open-mouthed and motionless. This largest freshwater turtle is found in rivers, streams and swamps in North America.

In the middle of the reefs, the nearly undetectable anglerfish, here *Rycherus filamentosus*, fish with bait. Each species has a lure-like shape of its own, resembling a small fish, a shrimp or a worm and which, in case of loss, regenerates, more or less according to the original, in four to six months. Once the "prey" has been spotted, the anglerfish shakes its bait, causing it to first make large movements visible from afar and then smaller and faster motions. The victim is approaching. The predator adjusts its position... When its future meal is at the right distance, the predator opens its huge mouth. The depression thereby created sucks the dinner into the esophagus of the anglerfish, whose stomach can contain a fish twice its own size!

Following spread
In addition to its lure, the striated frogfish has a chemical weapon. This species lives at depths of up to 650 feet (200 m), where visibility is poor. Its lure emits pheromones to attract victims over long distances. The striated frogfish positions itself against the currents that carry away the chemicals from its trap. By following these scents, the future victims soon face the lure... You know the rest!

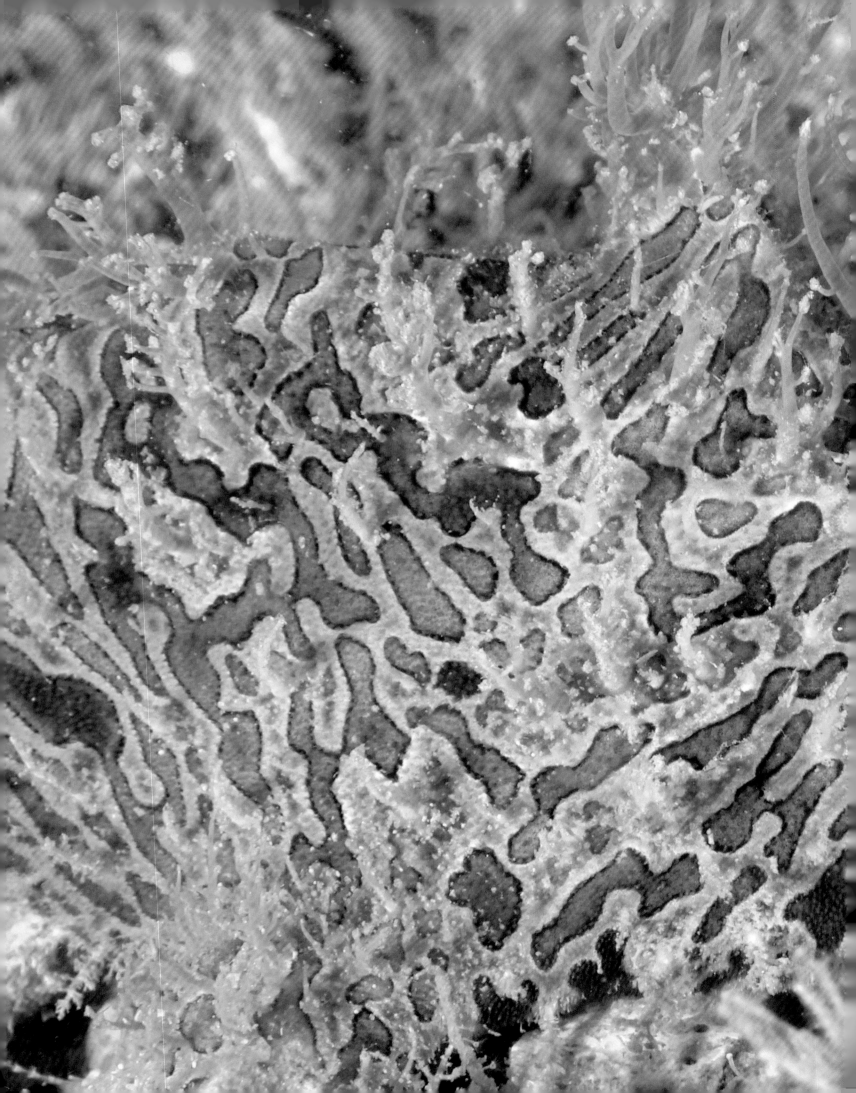

CONCLUSION

Thanks to the journey we have just been on and the surprising animals we have discovered, we now have a better understanding of the important role camouflage plays in nature, and we also understand the difficulty of attempting to clearly and precisely define and categorize it.

It was a subject of wonder and questioning for early evolutionists, but it mostly remained a field of secondary, even anecdotal, research. It was not until the 2000s that several teams of scientists around the world began to earnestly look into its mechanisms. It is now as much about perceiving the evolutionary mechanisms that allow an insect to look like a plant as understanding how animal cognition intervenes in the process of camouflage. What perception does an animal have of its own mimicry? How does camouflage get introduced into an environment? Experiments have shown that animals' ability to change their color and their decisions about where to position themselves are not based solely on what they know about their environment. Thus the caterpillar of the early thorn moth adapts its choice of branch to its hunger and the risk of predation incurred, while the crab *Tiarinia cornigera* covers its shell only if the presence of predators is significant. Today, scientists do not know anything about the mechanisms behind these kinds of decisions. Camouflage and its corollary, mimicry, constitute a vast field for scientific exploration. But is it perhaps already too late? Through the many examples in this book, we find that it is among tropical insects and reef dwellers that the diversity of camouflage is most widely expressed. However, these environments are the most endangered on the planet. Every big tree that collapses for the benefit of multinational corporations and each piece of coral that is bleached due to global warming carries with it a bit of the magic of camouflage.

Nature seems to have no limit; it is able to imagine anything. This petiole is obviously not a petiole. It's a spider. Spiders of the genus *Poltys* (here, *Poltys elevatus*) have specialized in the copying of plant forms. The only spider of this genus, present across Asia, Africa and Oceania, that imitates a leaf was discovered in the 2010s. These spiders remain motionless all day, depending on their perfect mimicry to go unnoticed. As night falls, they weave a web that is destroyed every morning. Neither seen nor known of! Now isn't that the art of the champions of camouflage?

PHOTO CREDITS

J.W. Alker/imageBROKER: 100
Ingo Arndt/Minden Pictures: 74-75, 145
Fabrice Audier: 32-33
Franco Banfi/Photoshot: 90-91
Fred Bavendam/Minden Pictures: 21, 114-115, 155
Nicky Bay/Science Photo Library: 123
Paul Bertner/Minden Pictures: 34-35, 140, 141
Husni Che Ngah: 124-125
Brandon Cole: 49
Stephen Dalton/Photoshot: 82, 118-119
Ethan Daniels/WaterFrame: 153
Steve De Neef/Visual and Written: 20
Frank Deschandol & Philippe Sabine: 151
Reinhard Dirscherl: 23
Bill Draker/imageBROKER: 85
Arjen Drost/Buiten-Beeld: 39
Alain Even: 84
Gianpiero Ferrari/Frank Lane Picture Agency: 78-79
David Fleetham/Visuals Unlimited/Science Photo Library: 46-47
Michael & Patricia Fogden/Minden Pictures: 108-109
Mathieu Foulquié: 52-53
Atsuo Fujimaru/Oasis: 83
Borut Furlan/WaterFrame: 101, 102, 127
Bruno Guénard: 6, 106, 120
Adrian Hepworth/Photoshot: 111
Daniel Heuclin: 154
Stephen Krasemann/Photoshot: 139
Jean Lecomte: 48
Chien Lee/Minden Pictures: 28, 65, 122, 146, 159
Jean-Claude Malausa: 121
Thomas Marent/Minden Pictures: 15, 29, 44-45, 58-59, 62-63, 112
Thomas Marent/Visuals Unlimited/Science Photo Library: 8-9, 18
Colin Marshall/Frank Lane Picture Agency: 104-105
Quentin Martinez: 60-61
Chris Mattison/Photoshot: 10, 13
Fabien Michenet: 103
Mark Moffett/Minden Pictures: 136, 142-143, 144
Robin Monchâtre: 70
Piotr Naskrecki/Minden Pictures: 31, 116-117, 147, 148
Chris Newbert/Minden Pictures: 24-25, 126

Josef Niedermeier/imageBROKER: 43
Gerald Nowak/WaterFrame: 89
Rolf Nussbaumer/imageBROKER: 80-81
Matthew Oldfield/Science Photo Library: 107
Pete Oxford/Minden Pictures: 27, 66-67, 68-69, 72-73, 77, 86-87, 132-133
Andrew Parkinson/Frank Lane Picture Agency: 4-5
Benoît Personnaz: 19
Rod Planck/Photoshot: 38
Wolfgang Poelzer/WaterFrame: 94-95
Michel Poinsignon: 16-17
Eric Polidori: 104
Mike Powles/Frank Lane Picture Agency: 42
Daniela Preissler-Dirscherl/WaterFrame: 22, 50-51
Norbert Probst/imageBROKER: 100
Michel Rauch: 11, 71
Jacques Rosès: 26
Jany Sauvanet/Photoshot: 135
Malcolm Schuyl/Frank Lane Picture Agency: 149
Scubazoo/Science Photo Library: 96-97
Gérard Soury: 92
Robert Thompson/Photoshot: 131
Francesco Tomasinelli/Photo Researchers: 56, 64, 113
Geoff Trinder/Ardea: 30
Michael Turco: 138
Yves Vallier: 93
Markus Varesvuo: 37
Jean Venot: 25
Alan J. S. Weaving/Ardea: 110
Terry Whittaker/Frank Lane Picture Agency: 40-41
Birgitte Wilms/Minden Pictures: 128-129, 137
Norbert Wu: 134
Norbert Wu/Minden Pictures: 57, 98-99, 156-157
ZSSD/Minden Pictures: 54-55

Front cover: Thomas Marent/Minden Pictures
Back cover: top left, Reinhard Dirscherl; top right, Terry Whittaker/FLPA - Frank Lane Picture Agency; bottom left, Chris Mattison/Photoshot; bottom right, Thomas Marent/Minden Pictures